Android
程序设计

主　编　黄　兴　张赵辉　张佳乐
副主编　龙运明　赵雪章

北京希望电子出版社
Beijing Hope Electronic Press
www.bhp.com.cn

内容简介

本书深入浅出地介绍了 Android 应用开发的知识技能，并通过大量实例使读者更好地理解和掌握在 Android 程序设计中的各知识点和在实际开发中的应用。本书共 9 章，内容包含了 Android 程序设计基础、界面布局、菜单栏和对话框、常用控件、多线程与事件处理机制、数据存储机制、Intent 和 ContentProvider、多媒体开发和电话 API、基于桌面组件的开发等知识。

本书内容翔实，实例丰富，图文并茂，基本覆盖了 Android 应用开发的各个方面。本书可用作本科院校软件类、计算机类和通信类等专业的教材，也适合作为相关培训学校的培训教材及从事 Android 移动应用开发人员的参考书籍。

图书在版编目（CIP）数据

Android 程序设计 / 黄兴, 张赵辉, 张佳乐主编.

北京：北京希望电子出版社, 2024.9.

ISBN 978-7-83002-860-2

Ⅰ. TN929.53

中国国家版本馆 CIP 数据核字第 20245JR279 号

出版：北京希望电子出版社	封面：袁　野
地址：北京市海淀区中关村大街 22 号	编辑：周卓琳
中科大厦 A 座 10 层	校对：付寒冰
邮编：100190	开本：787 mm×1092 mm　1/16
网址：www.bhp.com.cn	印张：15.5
电话：010-82620818（总机）转发行部	字数：368 千字
010-82626237（邮购）	印刷：北京昌联印刷有限公司
经销：各地新华书店	版次：2024 年 12 月 1 版 1 次印刷

定价：55.00 元

前 言
PREFACE

　　当下，智能手机的功能越来越多，为了实现这些功能，必须有一个好的开发平台。多年以来，Android系统得到了业界广泛的支持。Android是Google公司推出的基于Linux平台的开源手机操作系统，由于其开放性和优异的应用功能，成为目前最受欢迎的嵌入式操作系统之一，其发展势头势不可当。

　　移动终端的快速发展，使得Android系统应用需求激增，越来越多的编程人员甚至是在校生都加入到Android的开发阵营中。为了帮助开发者更快更好地进行Android的开发，笔者精心编写了此书。本书以满足实际需求为出发点，科学合理地安排知识结构，内容由浅入深、循序渐进地展开，具有较完善的知识体系，内容与当前Android技术的发展和应用水平相匹配。

　　本书共9章，建议总课时为42课时，具体安排如下：

章节	内容	理论教学	上机实训
第1章	Android程序设计基础	2课时	0课时
第2章	界面布局	2课时	2课时
第3章	菜单和对话框	2课时	2课时
第4章	常用控件	3课时	3课时
第5章	多线程与事件处理机制	2课时	2课时
第6章	数据存储机制	3课时	3课时
第7章	Intent和ContentProvider	3课时	3课时
第8章	多媒体开发和电话API	3课时	3课时
第9章	综合应用：基于桌面组件的开发	1课时	3课时

　　本书由重庆移通学院黄兴、黑龙江林业职业技术学院张赵辉、张家口职业技术学院张佳乐担任主编，由广西城市职业大学龙运明、佛山职业技术学院 赵雪章担任副主编。黄兴编写第1章、第2章和第7章部分内容，张赵辉编写第3章、第5章和6章部分内容，张佳乐编写第4章至第7章部分内容，龙运明和赵雪章编写第8章和第9章部分内容。本书在编写过程中，得到了企业的大力支持，在此一并表示感谢！本书在编写过程中力求严谨细致，但由于编者水平有限，疏漏之处在所难免，望广大读者批评指正。

编　者

2023年8月

目录 CONTENTS

第1章 Android程序设计基础

- 1.1 Android简介 ... 2
  - 1.1.1 Android的特点 ... 2
  - 1.1.2 Android系统架构 ... 2
- 1.2 Android应用程序的组成 ... 4
  - 1.2.1 R.java文件详解 ... 8
  - 1.2.2 组件标识符 ... 9
  - 1.2.3 AndroidManifest.xml文件详解 ... 10
- 1.3 Android应用程序的执行 ... 14
- 1.4 Android应用程序组件 ... 17
  - 1.4.1 Activity ... 17
  - 1.4.2 Service ... 17
  - 1.4.3 BroadcastReceiver ... 18
  - 1.4.4 ContentProvider ... 18
- 课后作业 ... 18

第2章 界面布局

- 2.1 View概述 ... 21
- 2.2 Android界面布局 ... 21
  - 2.2.1 LinearLayout（线性布局） ... 21
  - 2.2.2 RelativeLayout（相对布局） ... 24
  - 2.2.3 TableLayout（表格布局） ... 28
- 2.3 文本框及按钮控件 ... 31
- 课后作业 ... 35

第3章 菜单和对话框

- 3.1 选项菜单和子菜单 ... 38
  - 3.1.1 创建OptionsMenu菜单实例 ... 38
  - 3.1.2 监听菜单事件 ... 43

3.1.3 与菜单项关联的Activity的设置……44
3.2 上下文菜单……45
3.3 Android中的对话框……47
　　3.3.1 AlertDialog（提示对话框）……47
　　3.3.2 ProgressDialog（进度对话框）……53
3.4 提示信息……55
　　3.4.1 Toast……55
　　3.4.2 Notification……56

课后作业　57

第4章 常用控件

4.1 ImageButton控件……59
4.2 ImageView控件……61
4.3 单选按钮和复选框……63
　　4.3.1 单选按钮组和单选按钮的用法……63
　　4.3.2 复选框的用法……68
4.4 列表视图（ListView）……74
　　4.4.1 简单的列表视图……74
　　4.4.2 带标题的ListView列表……76
　　4.4.3 带图片的ListView列表……78
4.5 网格视图（GridView）……81

课后作业　85

第5章 多线程与事件处理机制

5.1 Android的多线程……88
　　5.1.1 多线程机制的优缺点……88
　　5.1.2 多线程的实现……90
5.2 事件处理机制……102
　　5.2.1 基于监听接口的事件处理……102
　　5.2.2 基于回调机制的事件处理……104
　　5.2.3 回调方法应用案例……106

课后作业　109

第6章 数据存储机制

- 6.1 Shared Preferences ... 111
- 6.2 存储数据到文件 ... 117
  - 6.2.1 实现过程 ... 117
  - 6.2.2 操作分析 ... 122
- 6.3 使用数据库存储数据 ... 126
  - 6.3.1 创建数据库帮助类DBAdapter ... 126
  - 6.3.2 添加数据到数据表books中 ... 130
  - 6.3.3 获取数据表中的所有记录 ... 131
  - 6.3.4 获取数据表中的某一条记录 ... 132
  - 6.3.5 更新数据表中的某一条记录 ... 134
  - 6.3.6 删除数据表中的某一条记录 ... 136
- 课后作业 ... 138

第7章 Intent和ContentProvider

- 7.1 Intent ... 141
  - 7.1.1 Intent的组成 ... 141
  - 7.1.2 Intent Filter ... 142
  - 7.1.3 Intent的解析 ... 144
  - 7.1.4 Intent的实现 ... 145
  - 7.1.5 Intent中传递数据 ... 150
  - 7.1.6 在Intent中传递复杂对象 ... 155
- 7.2 ContentProvider ... 160
  - 7.2.1 ContentProvider简介 ... 160
  - 7.2.2 Uri、UriMatcher、ContentUris和ContentResolver类简介 ... 161
  - 7.2.3 自定义ContentProvider ... 163
  - 7.2.4 系统ContentProvider ... 170
- 7.3 简单管理程序设计 ... 173
- 课后作业 ... 194

第8章 多媒体开发和电话API

- 8.1 多媒体开发 ... 197
  - 8.1.1 常见的多媒体格式 ... 197

	8.1.2	播放音频	198
	8.1.3	播放视频	202
	8.1.4	录制音频	203
	8.1.5	录制视频	207
8.2	使用电话API		217
	8.2.1	拨打电话	217
	8.2.2	发送SMS	217
	8.2.3	接收SMS	220

课后作业 222

第9章 综合应用：基于桌面组件的开发

9.1	桌面快捷方式介绍	225
9.2	桌面组件——Widget	226
	9.2.1 AppWidget框架类	226
	9.2.2 AppWidget的简单例子：Hello AppWidget	227
9.3	桌面天气预报程序设计	232

参考文献 240

第 1 章

Android 程序设计基础

── 内容概要 ──

Android 是一种基于 Linux 的开放源代码的操作系统，主要用于移动设备，如智能手机和平板电脑，它由 Google 公司和开放手机联盟引领及开发。Android 操作系统由于其良好的性价比、开放的环境、友好的人机交互接口等优点使得其获得了极高的市场占有率。本章将从其特点讲起，重点介绍 Android 系统架构、应用程序组成、应用程序组件等内容。

微信扫码

- 配套资源
- 入门精讲
- 项目实战
- 日志记录

1.1 Android简介

Android是一种基于Linux的开放源代码的操作系统，主要用于移动设备，如智能手机和平板电脑。Android操作系统最初由Andy Rubin开发，主要用于支持手机，2005年8月被Google收购。2007年11月，Google与84家硬件制造商、软件开发商及电信运营商组建开放手机联盟，共同研发改良Android系统。随后Google以Apache开源许可证的授权方式，发布了Android的源代码。第1款Android智能手机发布于2008年10月，随后逐渐扩展到平板电脑及其他领域，如电视、数码相机、游戏机等。

1.1.1 Android的特点

Android支持3D加速图形引擎，也支持SQLite数据库等。如果开发者熟悉Java编程或者其他任何种类的OOP开发，则可以使用用户接口（User Interface，UI）开发程序。Android允许使用UI开发，而且还支持以XML为基础的UI布局。XML UI布局对普通桌面开发者是一个非常新的概念。Android的开发者可以将自己的应用程序和Google提供的谷歌地图或谷歌搜索等绑定在一起。例如，写一个应用程序实现在谷歌地图上显示来电者的位置，或者实现存储搜索结果到联系人中。

与其他智能手机操作系统相比，Android具有以下优点。

1. 开放性

Google与开放手机联盟合作开发了Android。Google通过与运营商、设备制造商、软件开发商及其他有关各方结成深层次的合作伙伴关系，通过建立标准化、开放式的移动电话软件平台，在移动产业内形成了一个开放式的生态系统。

2. 应用程序无界限

Android上的应用程序可以通过标准应用程序接口（Application Programming Interface，API）访问核心移动设备，从而实现其特定的功能。通过互联网，这些应用程序又可供其他应用程序使用。

3. 应用平等

移动设备商的应用程序可以被替换或扩展，即使拨号程序或主屏幕这样的核心组件也是可以被替换的。

4. 快速方便的应用开发

Android平台为开发人员提供了大量的使用库和工具，开发人员可以利用它们快速地创建自己的应用。

1.1.2 Android系统架构

Android的优点取决于Android优秀的体系架构。Android的系统架构和其操作系统一样，采用了分层架构的思想,如图1-1所示。整个架构由应用程序层、应用程序框架层、Android运行时、系统库和Linux内核层构成。

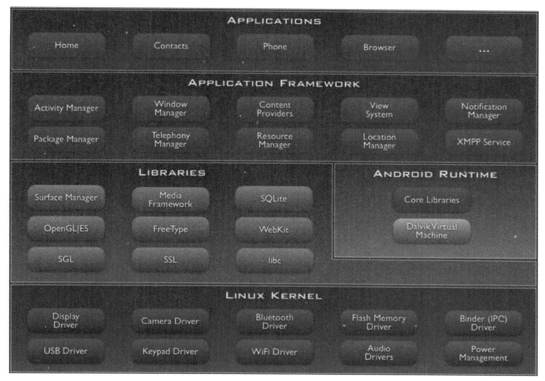

图 1-1 Android 系统架构

1. 应用程序层

Android平台默认包含了一系列的核心应用程序,包括电子邮件、短信、日历、地图、浏览器、联系人管理程序等,这些应用程序都是用Java语言编写并运行在虚拟机上的。当然,作为程序员也可以用自己写的程序替换Android提供的应用程序,这就需要应用程序框架层来保证了。

2. 应用程序框架层

这一层是进行Android开发的基础,开发人员可以使用这些框架开发自己的应用程序,这样便简化了程序开发的架构设计,但是必须遵守其框架的开发原则。应用程序框架层包含了视图系统、内容提供器、资源管理器、通知管理器、活动管理器、窗口管理器、电话管理器、包管理器等部分。

3. Android运行时

Android虽然采用Java语言开发应用程序,但却不使用J2ME执行Java程序,而是用Android自有的Android运行时来执行。Android运行时包括核心库和Dalvik虚拟机两部分。

核心库包含两部分内容,一部分提供Java编程语言核心库的大多数功能,另一部分为

Android的核心库。与标准的Java不同，Android不是用一个Dalvik虚拟机来同时执行多个Android应用程序，而是每个Android应用程序都用一个自有的Dalvik虚拟机来执行。

Dalvik虚拟机是一种基于寄存器的Java虚拟机，它是专为移动设备而设计的。Dalvik虚拟机在设计时就已经设想用最少的内存资源来执行，并支持同时执行多个虚拟机的特性。在设计方面，Dalvik虚拟机有许多地方参考了Java虚拟机，但是Dalvik虚拟机执行的中间码并非是Java虚拟机执行的Java字节码，同时它也不直接执行Java的类，而是依靠转换工具dx将Java字节码转换为Dalvik虚拟机执行时特有的dex（Dalvik Excutable）格式。

4. 系统库

应用程序框架层是贴近于应用程序的软件组件服务，而更底层则是Android的库函数（C/C++编写），这一部分是应用程序框架的支撑。这一层主要包括以下功能。

- **Surface Manager**：同时执行多个应用程序时，它负责管理显示与存取操作间的互动，另外也负责将2D绘图与3D绘图进行显示上的合成。
- **WebKit**：它是一套网页浏览器的软件引擎，该引擎的功能不仅可供Android内建的网页浏览器所调用，也可以提供内嵌性网页的显示效果。
- **SGL**：提供Android在2D绘图方面的绘图引擎。
- **OpenGL ES**：Android是依据OpenGL ES 1.0 API标准实现其3D绘图函数库的，该函数库可以用软件方式执行，也可以用硬件加速方式执行，其中3D软件光栅处理方面已进行高度优化。
- **FreeType**：提供点阵字、向量字的描绘显示。
- **媒体框架**：提供了对各种音频、视频的支持。Android支持多种音频、视频、静态图像格式，如MPEG-4、H.264、MP3、AAC、ARM、JPG、PNG、GIF等。
- **SQLite**：这是一套轻量级的数据库引擎，可供其他应用程序调用。
- **Libc**：提供针对移动设备优化的C库。

5. Linux内核层

Android平台的一个主要优点就是开放性，采用Linux内核则是Android平台开放性的基础。Linux内核层为软件层和硬件层建立了一个抽象层，使得应用开发人员无须关心硬件细节。不过对手机开发商而言，如果想要Android平台在自己的硬件平台上运行，就必须对Linux内核层进行修改，通常要做的工作就是为自己的硬件编写驱动程序。

1.2 Android应用程序的组成

在Android Studio中，一个Android应用程序项目包含有相对比较复杂的目录结构和文件，如资源文件、功能清单文件AndroidManifest.xml、Activity、Service、BroadcastReceiver、ContentProvider等，还需要通过Intent进行通信。刚开始接触这么复杂的目录结构和文件，可能会产生一些疑惑。这里将从程序的目录结构入手进行分析，Android Studio中的目录层次结构如图1-2所示。

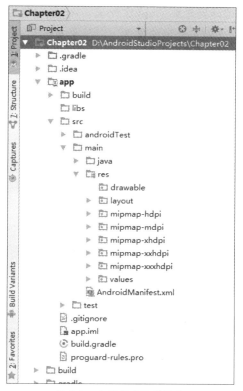

图 1-2　应用程序目录结构

如果读者以前使用过 Eclipse，在此需要区分清楚 Project 和 Module 这两个概念。在 Android Studio 中，Project 的真实含义是工作空间，相当于 Eclipse 中的 Workspace；Module 为一个具体的项目，相当于 Eclipse 中的 Project；一个 Project 可以包含多个 Module。Project 中主要的文件夹功能介绍如下。

- **.gradle 文件夹**：构建工具系统的 jar 和 wrapper 等，jar 的功能是告诉 Android Studio 如何与系统安装的 gradle 构建联系。
- **.idea 文件夹**：Android Studio 生成的工程配置文件，类似 Eclipse 的 project.properties。
- **app 文件夹**：Android Studio 创建的工程中的一个 Module。app 文件夹中包含一些子文件夹。

app 文件夹中包含的主要文件夹的功能介绍如下。

- **build**：构建目录，该目录不用开发人员维护。
- **libs**：存放依赖的包。
- **src\androidTest**：专门存放测试类，是运行在 emulator 和 device 上的测试方式，测试用例中所有的行为都是经过 Android framework 验证的。
- **src\test**：专门存放测试类，是脱离 emulator 和 device 独立运行在虚拟机上的测试方式。
- **src\main\java**：专门存放 Java 源代码的包文件。
- **src\main\res**：资源目录，用于存放工程中的资源文件。该目录可以存放一些图标文件、界面文件、布局文件、应用中用到的文字信息等。

程序编译后，资源会被编译到最终的apk文件里。Android创建了一个被称为R的类，这样在Java代码中可以通过它关联到对应的资源文件。以下是对res的子目录的详细说明。

- **AndroidManifest.xml**：该文件是Android项目的系统清单文件，用于控制Android应用的名称、图标、访问权限等整体属性。
- **drawable**：用于存放各种图像文件，如.png、jpg、gif等，除此之外还有一些其他的drawable类型的XML文件。
- **mipmap-hdpi**：用于存放高分辨率文件，一般是把图片放在这里。
- **mipmap-mdpi**：用于存放中等分辨率文件，很少使用，除非兼容的手机很旧。
- **mipmap-xhdpi**：用于存放超高分辨率文件，手机屏幕的材质越来越好，估计以后会慢慢往这里过渡。
- **mipmap-xxhdpi**：用于存放超超高分辨率文件，它在高端机上有所体现。
- **mipmap-xxxhdpi**：用于存放超超超高分辨率文件，它在高端机上有所体现。

drawable和mipmap区别其实不大，只是使用mipmap会在图片缩放上提供一定的性能优化，系统会根据屏幕分辨率来选择hdpi、mdpi、xmdpi、xxhdpi、xxxhdpi下的对应图片，所以，解压apk后可以看到上述目录中存在同一名称的图片，在5个文件夹中都有，只是大小和像素不一样而已。

- **res\values**：用于存放值资源，该文件夹中常放的文件有strings.xml和styles.xml。

（1）"strings.xml"文件用于定义字符串和数值，在Activity中可以使用getResources().getString(resourceId)或getResources().getText(resourceId)取得指定的资源。打开app项目的"strings.xml"文件，可以看到如下内容。

```
<resources>
    <string name="app_name">Chapter02</string>
</resources>
```

每个string标签声明一个字符串，name属性指定其引用名。为什么需要把应用中出现的文字单独放在"string.xml"文件中呢？原因有如下两点。

一是为了方便应用的国际化。Android建议将在屏幕上显示的文字定义在"strings.xml"中，如果以后需要进行国际化，例如，开发的应用原来是面向国内用户的，自然要在屏幕上使用中文，如今要让应用走向世界，如打入日本市场，当然就需要在手机屏幕上显示日语。如果没有把文字信息定义在"strings.xml"中，就需要修改程序内容了。但是，如果把所有屏幕上出现的文字信息都集中存放在"strings.xml"文件之后，那么就只需要再提供一个"strings.xml"文件，把其中的汉字信息都修改为日语即可。再运行程序时，Android操作系统会根据用户手机的语言环境和国家自动选择相应的"strings.xml"文件，这时手机界面就会显示出日语。这样做将使应用程序的国际化实现起来非常方便。二是为了减少应用的体积，降低数据冗余。假设在应用中要使用"我们一直在努力"这段文字10 000次，如果不将"我们一直在努力"定义在"strings.xml"文件中，而是在每次使用时直接写上这几个字，这样程序中将有

70 000个字，这70 000个字将占136 KB的空间。由于手机CPU的处理能力及内存是有限的，136 KB对手机程序来说是个不小的空间，做手机应用的原则之一就是"能省内存就省内存"，而如果将这几个字定义在"strings.xml"中，在每次使用到的地方通过Resources类来引用该文字，将只需占用14 B的空间，因此，这对降低应用的体积效果是非常明显的。

（2）"styles.xml"文件用于定义样式。打开本项目的"styles.xml"文件，内容如下。

```xml
<resources>
    <!-- Base application theme. -->
    <style name="AppTheme" parent="Theme.AppCompat.Light.DarkActionBar">
        <!-- Customize your theme here. -->
        <item name="colorPrimary">@color/colorPrimary</item>
        <item name="colorPrimaryDark">@color/colorPrimaryDark</item>
        <item name="colorAccent">@color/colorAccent</item>
    </style>
</resources>
```

需要注意的是，Android中的资源文件不要以数字作为文件名，这样会导致错误。

● res\layout：用于存放布局文件。

这里的布局文件是ADT默认自动创建的"activity_main.xml"文件。通常，可以用两种方式——Graphical Layout或者XML清单显示其中的内容，在Android Studio中，这两种查看方式可以随意切换。双击打开此XML文件，内容如下。

```xml
<?xml version="1.0" encoding="utf-8"?>
<android.support.constraint.ConstraintLayout xmlns:android="http://schemas.android.com/apk/res/android"
    xmlns:app="http://schemas.android.com/apk/res-auto"
    xmlns:tools="http://schemas.android.com/tools"
    android:layout_width="match_parent"
    android:layout_height="match_parent"
    tools:context="cn.androidstudy.chapter02.MainActivity">
    <TextView
        android:layout_width="wrap_content"
        android:layout_height="wrap_content"
        android:text="Hello World!"
        app:layout_constraintBottom_toBottomOf="parent"
        app:layout_constraintLeft_toLeftOf="parent"
        app:layout_constraintRight_toRightOf="parent"
        app:layout_constraintTop_toTopOf="parent" />
</android.support.constraint.ConstraintLayout>
```

与在网页中布局使用HTML文件一样，Android在XML文件中使用XML元素来设定屏幕布局。布局文件保存在工程的res\layout目录下，每个文件包含整个屏幕或部分屏幕，它被Android资源编辑器编译，编译时会被编译进一个视图资源，还可以将它传递给Activity.setContentView或被其他布局文件引用。

■1.2.1 R.java文件详解

Android创建了一个被称为R的类，在Java代码中可以通过它关联到对应的资源文件。"R.java"文件的位置如图1-3所示。

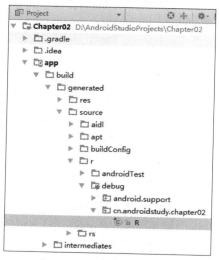

图1-3 "R.java"文件的位置

"R.java"文件中有一系列静态内部类，每个静态内部类分别对应一种资源，如layout静态内部类对应layout中的界面文件；每个静态内部类中的静态常量定义一条资源标识符，如"public static final int activity_main= 0x7f04001b;"对应的是layout目录下的"activity_main.xml"文件。具体的对应关系如图1-4所示。

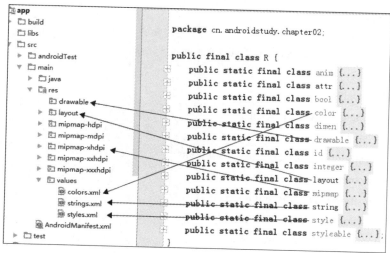

图1-4 "R.java"中资源的对应关系图

"R.java"文件中内容的来源，也即当开发者在res目录下任何一个子目录中添加相应类型的文件之后，会在"R.java"文件中相应的匿名内部类当中自动生成一条静态int类型的常量，对添加的文件进行"索引"。如果在layout目录下添加一个新的界面布局，那么在R.java中的public static final class layout类中也会添加相应的静态int类型常量。相反，当在res目录下删除任何一个文件，其在"R.java"中对应的记录也会被自动删除。例如，在"strings.xml"中添加一条记录，在"R.java"的string内部类中也会自动增加一条记录。"R.java"文件给开发程序带来很大的方便，例如，在程序中使用"public static final int ic_launcher=0x7f020000;"就可以找到其对应的ic_launcher这幅图片。"R.java"文件除了有自动标识资源的"索引"功能之外，还有另一个主要的功能，即当res目录中的某个资源在应用中没有被使用到，则在该应用被编译的时候，系统就不会把对应的资源编译到该应用的apk包中，这样可以节省Android手机的资源。

■1.2.2 组件标识符

"R.java"文件具有索引的作用，通过它可以检索到应用中需要使用的资源。如何通过"R.java"文件引用所需要的资源呢？具体有以下3种方式。

（1）在Java程序中可以按照Java的语法来引用。

①R.resource_type.resource_name。

需要注意的是，resource_name不需要文件的后缀名。

例如，上面的"ic_launcher.png"文件的资源标识符可以通过如下方式获取。

R.mipmap.ic_launcher

②android.R.resource_type.resource_name。

Android系统本身自带了很多资源，这些资源也都是可以进行引用的，只是引用时需要在前面加上"android."以声明该资源来自Android系统。

（2）在XML文件中引用资源。

这种方式引用资源的语法如下。

①@[package:]type/name。

注意，使用自己包下的资源可以省略"package:"。

XML文件（如"activity_main.xml"和"AndroidManifest.xml"文件）是通过"@mipmap/ic_launcher"的方式获取资源，其中"@"代表的是R.java类，"mipmap"代表R.java中的静态内部类mipmap，"/ic_launcher"代表静态内部类mipmap中的静态属性ic_launcher，而该属性是指向res目录下的mipmap-*dpi（*为通配符，表示0个或多个字符）目录中的"ic_launcher.png"图标文件。

其他类型文件的引用也类似。凡是在"R.java"文件中定义的资源都可以通过"@Static_inner_classes_name/resourse_name"的方式获取，如"@id/button""@string/app_name"等。

②如果访问的是Android系统自带的文件，则要在引用方式的前面添上包名"android:"。

例如，android:textColor="@android:color/red"。

(3)"@+id/string_name"表达式方式。

在布局文件当中需要为一些组件添加id属性作为标识,可以使用的表达式形式为"@+id/string_name",其中,"+"表示在"R.java"文件中的名为id的内部类中添加一条记录。例如,"@+id/button"的含义是在"R.java"文件中的id这个静态内部类添加一个名为button的常量,该常量就是该资源的标识符。如果id这个静态内部类不存在,则会先生成它。通过该方式生成的资源标识符,仍然可以用"@id/string_name"这样的方式引用。示例代码片段如下。

```
<RelativeLayout
    android:layout_width = "fill_parent"
    android:layout_height = "wrap_content"
>
<Button
    android:layout_width = "wrap_content"
    android:layout_height = "wrap_content"
    android:text = "@string/cancel_button"
    android:layout_alignParentRight = "true"
    android:id = "@+id/cancel" />
<Button
    android:layout_width = "wrap_content"
    android:layout_height = "wrap_content"
    android:layout_toLeftOf = "@id/cancel"
    android:layout_alignTop = "@id/cancel"
    android:text = "@string/ok_button" />
</ RelativeLayout >
```

其中,android:id="@+id/cancel"将其所在的Button标识为cancel,在第2个Button中通过"@id/cancle"对第1个Button进行引用。

1.2.3 AndroidManifest.xml文件详解

每个应用程序都有一个功能清单文件"AndroidManifest.xml",这个清单文件给Android系统提供了关于这个应用程序的基本信息,系统在运行任何程序代码之前必须知道这些信息。用户开发Activity、Broadcast、Service之后都要在"AndroidManifest.xml"中进行定义。另外,如果要使用系统自带的服务,如拨号服务、应用安装服务、GPRS服务等,也都必须在"AndroidManifest.xml"中声明权限。

"AndroidManifest.xml"主要具有以下功能。

- 命名应用程序的Java应用包,这个包名用于唯一标识应用程序。
- 声明应用程序所必须具备的权限,用于访问受保护的部分API,以及和其他应用程序交互。
- 声明应用程序其他的必备权限,用于组件之间的交互。

- 列举测试设备Instrumentation类，用于提供应用程序运行时所需的环境配置及其他信息，这些声明只在程序开发和测试阶段存在，发布前将被删除。
- 声明应用程序所要求的Android API的最低版本级别。
- 列举应用程序需要链接的库。

下面以一个项目的功能清单文件为例进行说明，该"Android Manifest.xml"的内容如下。

```xml
<?xml version="1.0" encoding="utf-8"?>
<manifest xmlns:android="http://schemas.android.com/apk/res/android"
    package="cn.androidstudy.chapter02">
    <application
        android:allowBackup="true"
        android:icon="@mipmap/ic_launcher"
        android:label="@string/app_name"
        android:roundIcon="@mipmap/ic_launcher_round"
        android:supportsRtl="true"
        android:theme="@style/AppTheme">
        <activity android:name=".MainActivity">
            <intent-filter>
                <action android:name="android.intent.action.MAIN" />
                <category android:name="android.intent.category.LAUNCHER" />
            </intent-filter>
        </activity>
    </application>
</manifest>
```

下面将对各个标签进行详细说明。

（1）<manifest>元素。

```xml
<manifest xmlns:android="http://schemas.android.com/apk/res/android"
    package="cn.androidstudy.chapter02">
```

该元素是"AndroidManifest.xml"文件的根元素，是必须要有的。其中，根据XML文件的语法，"xmlns:android"用于指定该文件的命名空间，即使用"http://schemas.android.com/apk/res/android"所指向的一个文件；"package"属性用于指定Android应用所在的包。

（2）<application>元素。

```xml
<application
    android:allowBackup="true"
    android:icon="@mipmap/ic_launcher"
```

```xml
        android:label="@string/app_name"
        android:roundIcon="@mipmap/ic_launcher_round"
        android:supportsRtl="true"
        android:theme="@style/AppTheme">
        <activity android:name=".MainActivity">
            <intent-filter>
                <action android:name="android.intent.action.MAIN" />
                <category android:name="android.intent.category.LAUNCHER" />
            </intent-filter>
        </activity>
    </application>
```

　　<application>是非常重要的一个元素，用户开发的许多组件都会在该元素下定义。该元素为必选元素。<application>的icon属性用于设定应用的图标。<application>的label属性用于设定应用的名称，指定其属性值所用的表达式"@string/app_name"的含义与其下一行的表达式"@mipmap/ic_launcher"一样，同样是指向"R.java"文件string静态内部类中的app_name属性所指向的资源。在这里它指向的是"strings.xml"文件中的一条记录"app_name"，其值为"Chapter02"，因此，这种表达方式等价于android:label="Chapter02"。

　　（3）<activity>元素。

　　<activity>元素的作用是注册一个Activity信息，当程序开发人员在创建"Chapter02"项目时指定Activity的name属性为"MainActivity"，在生成项目时就会自动创建一个Activity，名称为"MainActivity.java"。Activity在Android中属于组件，它需要在功能清单文件中进行配置。<activity>元素的name属性指定的是该Activity的类名。<activity>元素的label属性表示Activity所代表的屏幕的标题，其属性值的表达式在上面已经介绍过了，不再赘述。该属性值在AVD运行程序到该Activity所代表的界面时，会在标题上显示该值。

　　（4）<intent-filter>元素。

　　intent-filter翻译成中文是"意图过滤器"。应用程序的核心组件（活动、服务和广播接收器）通过意图被激活，意图代表的是你要做的一件事情，代表你的目的。如果需要会启动这个组件的一个新实例，并传递给这个意图对象，Android会寻找一个合适的组件来响应这个意图。关于intent，后面会有详细的介绍，在此只需有这样的大致印象即可。

　　组件通过意图过滤器通告它们所具备的功能——能响应的意图类型。由于Android系统在启动一个组件前必须知道该组件能够处理哪些意图，因此，意图过滤器需要在manifest中以<intent-filter>元素指定。一个组件可以拥有多个过滤器，用于描述该组件所具有的不同能力。一个指定目标组件的显式意图将会激活那个指定的组件，此时意图过滤器不起作用。但一个没有指定目标的隐式意图只在它能够通过组件过滤器时才能激活该组件。

　　例如，示例中给出的过滤器为：

```
<intent-filter>
    <action android:name="android.intent.action.MAIN" />
    <category android:name="android.intent.category.LAUNCHER" />
</intent-filter>
```

以上过滤器是最常见的,它表明这个Activity将在应用程序加载器中显示,即用户在设备上看到的可供加载的应用程序列表。换句话说,这个Activity是应用程序的入口,是用户选择运行这个应用程序后所见到的第一个Activity。

(5) 权限Permissions。

本节给出的功能清单文件中并没有出现<Permissions>元素,但是Permission也是一个非常重要的节点,在后面的学习中会经常用到。Permission是指代码对设备上数据的访问限制,这个限制被引入用来保护可能因被误用而曲解或破坏用户体验的关键数据和代码,如拨号服务、短信服务等。每个许可被一个唯一的标签所标识,这个标签常常指出了受限的动作。

例如,申请发送短信服务的权限,需要在功能清单文件中添加如下语句。

```
<uses-permission android:name="android.permission.SEND_SMS"/>
```

一个功能最多只能被一个权限许可保护。如果一个应用程序需要访问一个需要特定权限的功能,它必须在<manifest>元素内使用<uses-permission>元素来声明这一点。这样,当应用程序安装到设备上之后,安装器可以通过检查签署应用程序认证的机构来决定是否授予请求的权限,在某些情况下,可能会询问用户。如果权限已被授予,那应用程序就能够访问受保护的功能特性。如果没有被授予,访问将失败,但不会给用户任何通知。因此,用户在使用一些系统服务(如拨号、短信、访问互联网、访问存储卡等)时一定要记得添加相应的权限,否则会出现一些难以预料的错误。

应用程序还可以通过权限许可来保护它自己的组件(活动、服务、广播接收器、内容提供者)。它可以利用Android已经定义(列在android.Manifest.permission里)或其他应用程序已声明的权限许可,或者定义自己的许可。一个新的许可要通过<permission>元素声明。例如,一个Activity可以用以下方式保护。

```
<manifest … >
    <permission android:name="com.example.project.DEBIT_ACCT" … />
    …
    <application … >
        <activity android:name="com.example.project.FreneticActivity" …
            android:permission="com.example.project.DEBIT_ACCT"
            … >
            …
```

```
        </activity>
    </application>
    ...
    <uses-permission android:name="com.example.project.DEBIT_ACCT" />
    ...
</manifest>
```

注意，在上面这个例子中，对DEBIT_ACCT的许可并非仅仅在<permission>元素中声明。如果该应用程序的其他组件要使用该组件，那么它同样需要在<uses-permission>元素里声明。

（6）库Libraries。

每个应用程序都链接到缺省的Android库，这个库包含了基础应用程序开发包（如活动、服务、意图、视图、按钮、应用程序、内容提供者等）。然而，一些包处于它们自己的库中。如果一个应用程序使用了其他开发包中的代码，则它必须显式的请求链接到它们，此时manifest必须包含一个单独的<uses-library>元素来命名每一个库，例如，在进行单元测试的时候需要引入其所需要的库。

代码片段如下。

```
<application android:icon="@drawable/icon" android:label="@string/app_name">
    <uses-library android:name="android.test.runner" />
</application>
```

1.3 Android应用程序的执行

经过环境配置和编写代码之后，程序的执行往往是比较令人期盼的。在熟悉了Android项目组成结构以及相关文件之后，下面将进一步对Android程序的具体编译和执行过程进行详细分析，带领读者深入Android的编程世界。

首先，要生成apk，Android程序的编译过程如图1-5所示。

（1）资源打包。使用aapt，即IDE中的资源打包工具（Android Asset Packaging Tool）将应用中的资源文件进行编译，这些资源文件包括"AndroidManifest.xml"文件、为Activity定义的XML文件等。在编译过程中也会产生一个"R.java"文件，这样就可以在Java代码中引用这些资源了。

（2）用aidl（Android Interface Definition Language）工具将项目中的所有.aidl接口转换成Java接口。

（3）编译项目中的所有Java代码，包括"R.java"和".aidl"文件，都会被Java编译器编译，然后输出.class文件。

（4）生成dex文件。使用Android Dalvik执行程序（Dalvik VM Executes）将上一步骤产生的.class文件转成Dalvik字节码，也就是.dex文件。同时，项目中包含的所有第三方类库和.class

第1章 Android程序设计基础

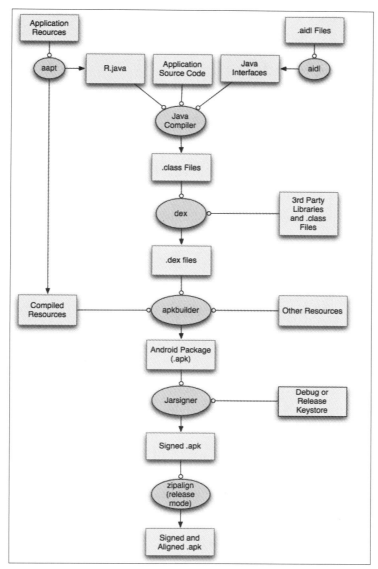

图 1-5　Android 程序的编译过程

文件也会被转换成.dex文件，这样方便下一步被打包成最终的.apk文件。

（5）生成apk文件。所有不能编译的资源（如图片等）、编译后的资源文件和.dex文件会被apkbuilder工具打包成一个.apk文件。

（6）签名apk文件，使用Jarsigner对apk文件进行签名。.apk文件被构建好之后，如果要把它安装到设备上，就必须用一个debug或者发行key来对这个apk文件签名。

（7）对齐apk，以使应用程序在设备上运行时减少内存消耗。对齐的应用在系统中执行时可以通过共享内存IPC读取资源，从而得到较高的性能。最后，如果应用程序已经被签名成为发行模式的apk文件，还需要使用Aipalign工具对.apk文件进行对齐优化，这样可以减少应用程序在设备上的内存消耗。

其次,将生成的apk文件安装到手机或者模拟器上。发布程序到手机上之后,当双击该应用的图标时,系统会将这个点击事件包装成一个Intent,该Intent包含两个参数:

{ action : "android.intent.action.MAIN",
  category : "android.intent.category.LAUNCHER" },

意图被传递给应用之后,Android将在应用的功能清单文件中去寻找与该意图匹配的意图过滤器,如果匹配成功,找到相匹配的意图过滤器所在的Activity元素,再根据<activity>元素的name属性寻找其对应的Activity类。接着Android操作系统创建该Activity类的实例对象,对象创建完成之后,会执行该类的onCreate方法,此OnCreate方法是通过重载其父类Activity的OnCreate方法而实现的。onCreate方法用于初始化Activity实例对象。例如,下面是HelloWorld.java类中onCreate方法的代码。

```
@Override
protected void onCreate(Bundle savedInstanceState) {
    super.onCreate(savedInstanceState);
    setContentView(R.layout.activity_main);
}
```

其中,super.onCreate(savedInstanceState)的作用是调用其父类Activity的OnCreate方法来实现对界面的绘制工作。在实现自己定义的Activity子类的OnCreate方法时一定要记得调用该方法,以确保能够绘制界面。

setContentView(R.layout.activity_main)的作用是加载一个界面。该方法中传入的参数是"R.layout.activity_main",其含义为R.java类中静态内部类layout的静态常量activity_main的值,而该值是一个指向res目录中layout子目录下"activity_main.xml"文件的标识符。因此,它代表显示activity_main所定义的画面。

Android程序执行的整个序列图如图1-6所示。

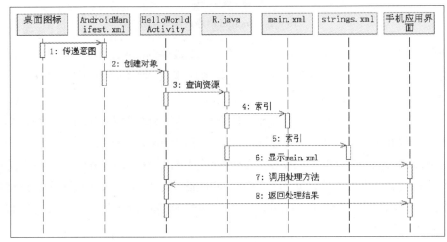

图1-6　Android应用执行的序列图

1.4 Android应用程序组件

Android程序中没有入口点（Main方法），取而代之的是一系列的应用程序组件，这些组件都可以单独实例化。本节将介绍Android支持的4种应用程序组件。应用程序对外共享功能一般也是通过这4种应用程序组件实现的。

1.4.1 Activity

Activity是Android的核心类，该类的全名是android.app.Activity。Activity相当于C/S程序中的窗体（form）或Web程序中的页面。每一个Activity都提供了一个可视化的区域，在这个区域可以放置各种Android控件，如按钮、图像、文本框等。

在Activity类中有一个onCreate事件方法，一般在该方法中对Activity进行初始化。通过setContentView方法可以将View绑定到Activity上。绑定后，Activity可以显示View上的控件。

一个带界面的Android应用程序可以由一个或多个Activity组成。至于这些Activity如何工作，或者它们之间有什么依赖关系，则完全取决于应用程序的业务逻辑。例如，一种典型的设计方案是使用一个Activity作为主Activity（相当于主窗体，程序启动时会首先显示这个Activity），在这个Activity中通过菜单、按钮等方式添加其他的Activity。在Android自带的程序中有很多是这种类型的。

每一个Activity都会有一个窗口，默认情况下，这个窗口是充满整个屏幕的。开发人员可以将窗口设置得比手机屏幕小，或者悬浮在其他窗口上面。Activity窗口中的可视化组件由View及其子类组成，这些组件按照XML布局文件中指定的位置在窗口中摆放。

1.4.2 Service

Service（服务）组件没有可视化的用户界面，但可以在后台运行。例如，当用户进行其他操作时，可以利用服务在后台播放音乐；或者在来电时，可以利用服务同时进行其他操作。Service类必须从android.app.Service继承。

以一个使用Service的简单例子说明，如在手机中，人们经常会使用播放音乐的软件，这类软件通常有循环播放或随机播放的功能。通常情况下，用户可能会一边放音乐，一边在手机上做其他的事，如聊天或看小说等。在这种情况下，用户一般不可能在一首音乐播放完后再回到播放音乐的软件界面去进行"重放"的操作（虽然在软件中可能会有相应的功能，如通过按钮或菜单进行控制），而是希望音乐能自动循环播放。因此，可以在播放音乐的软件中启动一个服务，由这个服务来控制音乐的循环播放，而且该服务对用户是完全透明的，用户完全感觉不到服务在后台运行；该服务甚至可以在音乐播放软件关闭的情况下，仍然播放后台背景音乐。

另外，其他的程序还可以与服务进行通信。当程序与服务连接成功后，就可以利用服务中共享出来的接口与服务进行通信了。例如，控制音乐播放的服务允许用户执行暂停、重放、停止播放等操作。

1.4.3 BroadcastReceiver

BroadcastReceiver（广播接收器）组件的唯一功能是接收广播动作，以及对广播动作做出响应。很多时候，广播动作是由系统发出的，如时区的变化、电池电量不足、收到短信等。除此之外，应用程序也可以发送广播动作，例如，通知其他程序，数据已经下载完毕，并且这些数据已经可以使用了。

一个应用程序可以有多个广播接收器，但所有的广播接收器类都需要继承android.content.BroadcastReceiver类。

广播接收器与服务一样，都没有用户接口，但在广播接收器中可以启动一个Activity来响应广播动作，如通过显示一个Activity对用户进行提醒等。当然，也可以采用其他方法或几种方法的组合来提醒用户，如闪屏、震动、响铃、播放音乐等。

1.4.4 ContentProvider

ContentProvider（内容提供者）组件可以为其他应用程序提供数据，这些数据可以保存在文件系统中。每一个内容提供者都是一个类，这些类都需要从android.content.ContentProvider类继承。

在ContentProvider类中定义了一系列的方法，通过这些方法可以使其他的应用程序获得内容提供者提供的数据。但在应用程序中不能直接调用这些方法，而是需要通过android.Content.ContentResolver类的方法来调用内容提供者类中提供的方法。

在Android系统中，很多内嵌的应用程序（如联系人、短信等）都提供了ContentProvider类。其他的应用程序通过这些ContentProvider类可以对系统内部的数据实现增、删、改等操作。例如，可以将指定电话号码的短信内容从系统数据库中删除，并将该短信内容加密保存在自己的数据库中，这些从系统数据库中删除短信的操作就需要通过内容提供者来完成。

课后作业

一、填空题

1. 目前，常见的智能手机操作系统有Harmony OS、_____、iOS等。

2. 与其他智能手机操作系统相比，Android操作系统具有以下的优点：_____、_____、_____、_____。

3. Android体系架构采用了软件叠层的技术，整个架构由_____、_____、_____、_____以及_____五层构成。

二、选择题

1. Android采用（　　）语言来开发、编写应用程序的。

　　A. Python

　　B. SQL

C. Java

D. C#

2. Android应用程序需要打包成（　　）文件格式在手机上安装运行。

A. .apk

B. .class

C. .xml

D. .dex

3. 中等分辨率的图像文件一般放在res目录的（　　）子目录下。

A. mipmap-hdpi

B. mipmap-mdpi

C. mipmap-xhdpi

D. mipmap-xxhdpi

4. （　　）是Android应用程序的表示层。

A. Service

B. Activity

C. BroadcastReceiver

D. ContentProvider

第 2 章

界面布局

内容概要

随着Android系统应用程序被广泛使用，设计一款美观、优雅、大气又实用的应用程序界面变得越来越重要，这将涉及到UI设计。本章主要讲解Android屏幕布局和常用基本控件的使用。

数字资源

【本章案例文件】："案例文件\第2章"目录下

- 配套资源
- 入门精讲
- 项目实战
- 日志记录

2.1　View概述

　　View（视图）是Android中最基本的一个类，Layout（布局）和Widget（控件）都是继承自View类。View对象是Android平台上表示用户界面的基本单元。视图是一个矩形区域，它负责这个区域里的绘制和事件处理。ViewGroup（视图组）是View的子类，它是一个容器，专门负责布局。视图组本身没有可绘制的元素。

2.2　Android界面布局

　　View一般可分为以下几种布局：LinearLayout（线性布局）、RelativeLayout（相对布局）、TableLayout（表格布局）、FrameLayout（单帧布局）和AbsoluteLayout（绝对布局）。本节主要学习线性布局、相对布局和表格布局，布局主要是在"res\layout\activity_main.xml"文件中编写。

■ 2.2.1　LinearLayout（线性布局）

　　线性布局是指子控件在该容器内的摆放方式，一般有垂直布局和水平布局两种。线性布局一般通过android:orientation和android:layout_weight两个属性来设置，属性的说明如表2-1所示。

表2-1　线性布局中两个重要的属性

xml属性	作用
android:orientation	设置布局的线性方向： horizontal：水平方向，从左到右 vertical：垂直方向，从上到下
android:layout_weight	设置控件占屏幕的比例，在垂直布局时，代表行距；水平布局时，代表列宽；weight值越大，控件所占屏幕的比例就越大

　　下面通过一个简单的案例来学习线性布局及其属性。

　　【案例2-1】：将Activity界面分成上下两部分，上部分占1/3，下部分占2/3。上部分用横向（水平）布局，放置3个Button；下部分用纵向（垂直）布局，也放置3个Button。线性布局效果图如图2-1所示。

　　1. 案例分析

　　要实现这样的布局必须要使用嵌套布局。

　　（1）最外层是一个垂直布局的线性布局。

　　（2）在最外层的线性布局中再嵌套两个（上、下）线性布局。

　　（3）上部分的线性布局使用水平布局，里面放3个Button。

　　（4）下部分的线性布局使用垂直布局，里面放3个Button。

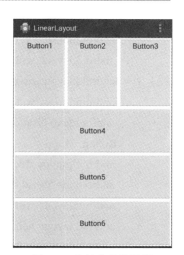

图2-1　线性布局效果图

2. 操作步骤

步骤 01 创建一个Android工程，命名为"LinearLayout"。

步骤 02 展开"Package Explorer"窗口中的"LinearLayout"项目，打开"res\layout\activity_main.xml"文件，修改并输入一些代码，代码如下。

```xml
<LinearLayout
    xmlns:android="http://schemas.android.com/apk/res/android"
    xmlns:tools="http://schemas.android.com/tools"
    android:layout_width="fill_parent"
    android:layout_height="fill_parent"
    android:orientation="vertical"
 >

<LinearLayout
    android:orientation="horizontal"
    android:layout_width="fill_parent"
    android:layout_height="fill_parent"
    android:layout_weight="2">
    <Button
      android:gravity="center_horizontal"
      android:id="@+id/button1"
      android:layout_width="wrap_content"
      android:layout_height="fill_parent"
      android:layout_weight="1"
      android:text="Button1" />
    <Button
      android:gravity="center_horizontal"
      android:id="@+id/button2"
      android:layout_width="wrap_content"
      android:layout_height="fill_parent"
      android:layout_weight="1"
      android:text="Button2" />
    <Button
      android:gravity="center_horizontal"
      android:id="@+id/button3"
      android:layout_width="wrap_content"
      android:layout_height="fill_parent"
      android:layout_weight="1"
```

```
            android:text="Button3" />
    </LinearLayout>

    <LinearLayout
        android:orientation="vertical"
        android:layout_width="fill_parent"
        android:layout_height="fill_parent"
        android:layout_weight="1" >
        <Button
            android:id="@+id/button4"
            android:layout_width="fill_parent"
            android:layout_height="wrap_content"
            android:layout_weight="1"
            android:text="Button4" />
        <Button
            android:id="@+id/button5"
            android:layout_width="fill_parent"
            android:layout_height="wrap_content"
            android:layout_weight="1"
            android:text="Button5" />
        <Button
            android:id="@+id/button6"
            android:layout_width="fill_parent"
            android:layout_height="wrap_content"
            android:layout_weight="1"
            android:text="Button6" />
    </LinearLayout>

</LinearLayout>
```

> **提示**：一般情况下，布局的代码是不加注释的。如果要加，可以加一些共用的属性，如MainActivity、Java等文件中的代码。一般情况下，关键的语句在代码中就注释了，不会单独拿出来注释的。

常用布局属性的说明如下。

- **android:layout_width**：用于设置控件的宽度。
- **android:layout_height**：用于设置控件的高度。
- **android:layout_weight**：用于设置控件占屏幕的比例。

- **android:id**：用于设置控件的ID。
- **android:text**：用于设置控件的显示文本。
- **android:gravity**：用于设置控件的基本位置。
- **android:background**：用于设置控件的背景。
- **android:textSize**：用于设置控件文字的大小。

2.2.2　RelativeLayout（相对布局）

相对布局是指一个容器内的子元素们通过彼此之间的位置来相互定位，或者与其父控件容器进行相互定位。

相对布局有一些重要的属性，这些属性一般分为4组。

第1组属性：用于设置控件与给定控件之间的关系和位置，其说明如表2-2所示。

表 2-2　确定控件之间的关系和位置的属性

属性	作用	备注
android:layout_above	将该控件置于给定ID的控件之上	属性值为某个给定控件的ID，例如：android:layout_above="@id/XXX"
android:layout_below	将该控件置于给定ID的控件之下	
android:layout_toLeftOf	将该控件置于给定ID的控件之左	
android:layout_toRightOf	将该控件置于给定ID的控件之右	

第2组属性：用于设置控件与控件之间的对齐方式，其说明如表2-3所示。

表 2-3　控件与控件之间对齐的属性

属性	作用	备注
android:layout_alignBaseline	该控件的baseline和给定ID控件的baseline对齐	属性值为某个给定控件的ID，例如：android:layout_above="@id/XXX"
android:layout_alignBottom	将该控件的底部边缘与给定ID控件的底部边缘对齐	
android:layout_alignLeft	将该控件的左侧边缘与给定ID控件的左侧边缘对齐	
android:layout_alignTop	将该控件的顶部边缘与给定ID控件的顶部边缘对齐	
android:layout_alignRight	将该控件的右侧边缘与给定ID控件的右侧边缘对齐	

第3组属性：用于设置控件与父控件之间的对齐方式，其说明如表2-4所示。

表 2-4　控件与父控件之间对齐的属性

属性	作用	备注
android:layout_alignParentBottom	该控件的底部与父控件的底部对齐	属性值为true或false，如果不指定，默认为false。属性表示的是控件自身的上下左右边缘与父控件的对应边缘是否对齐。因为控件是在父控件的内部，所以是内对齐
android:layout_alignParentLeft	该控件的左边缘与父控件的左边缘对齐	
android:layout_alignParentRight	该控件的右边缘与父控件的右边缘对齐	
android:layout_alignParentTop	该控件的顶部与父控件的顶部对齐	

第4组属性：用于设置控件的方向，其说明如表2-5所示。

表 2-5　控件方向的属性

属性	作用	备注
android:layout_centerHorizontal	该控件在其父控件范围内水平居中	属性值为true或false，如果不指定，默认为false。属性表示的是控件自身相对于父控件范围内的居中情况
android:layout_centerInparent	该控件在其父控件范围内垂直且水平居中	
android:layout_centerVertical	该控件在其父控件范围内垂直居中	

下面通过一个简单的案例来学习相对布局和设置控件的属性。

【案例2-2】：运用相对布局制作一张由9张小图片组合而成的大图片，效果如图2-2所示。

图 2-2　相对布局效果图

> 提示：使用9个TextView控件，在TextView控件中显示图片采用的是属性android:background。

1. 案例分析

要实现这样的布局需采用相对布局。

（1）导入项目中要用到的图片。

（2）在"activity_main.xml"文件中进行相对布局。

（3）在布局中放9个TextView控件，第1个控件放在界面中间，其他控件以第1个控件为参照物，放到相对应的位置。

2. 操作步骤

步骤01 创建一个Android工程，命名为"RelativeLayout"。

步骤02 导入9张图片并放到"res\drawable"目录对应的5个文件夹中。

步骤03 展开"Package Explorer"窗口中的"RelativeLayout"项目，打开"res\layout\activity_main.xml"文件，修改并输入一些代码，代码如下。

```xml
<RelativeLayout xmlns:android="http://schemas.android.com/apk/res/android"
  xmlns:tools="http://schemas.android.com/tools"
  android:layout_width="match_parent"
  android:layout_height="match_parent"
  android:paddingBottom="@dimen/activity_vertical_margin"
  android:paddingLeft="@dimen/activity_horizontal_margin"
  android:paddingRight="@dimen/activity_horizontal_margin"
  android:paddingTop="@dimen/activity_vertical_margin"
  tools:context=".MainActivity" >
  <TextView
    android:id="@+id/center"
    android:layout_width="wrap_content"
    android:layout_height="wrap_content"
    android:layout_centerInParent="true"
    android:background="@drawable/five"/>
  <TextView
    android:id="@+id/left"
    android:layout_width="wrap_content"
    android:layout_height="wrap_content"
    android:layout_above="@id/center"
    android:layout_toLeftOf="@id/center"
    android:background="@drawable/one" />
  <TextView
```

```
        android:id="@+id/right"
        android:layout_width="wrap_content"
        android:layout_height="wrap_content"
        android:layout_above="@id/center"
        android:layout_toRightOf="@id/center"
        android:background="@drawable/three" />"
    <TextView
        android:id="@+id/unright"
        android:layout_width="wrap_content"
        android:layout_height="wrap_content"
        android:layout_below="@id/center"
        android:layout_toLeftOf="@id/center"
        android:background="@drawable/seven" />
    <TextView
        android:layout_width="wrap_content"
        android:layout_height="wrap_content"
        android:layout_below="@id/center"
        android:layout_toRightOf="@id/center"
        android:background="@drawable/nine"/>
    <TextView
        android:layout_width="wrap_content"
        android:layout_height="wrap_content"
        android:layout_toRightOf="@id/center"
        android:layout_below="@id/right"
        android:background="@drawable/six" />"
    <TextView
        android:layout_width="wrap_content"
        android:layout_height="wrap_content"
        android:layout_toLeftOf="@id/center"
        android:layout_below="@+id/left"
        android:background="@drawable/four"/>
    <TextView
        android:layout_width="wrap_content"
        android:layout_height="wrap_content"
        android:layout_above="@id/center"
        android:layout_toRightOf="@+id/left"
        android:background="@drawable/two"/>
```

```
<TextView
    android:layout_width="wrap_content"
    android:layout_height="wrap_content"
    android:layout_below="@id/center"
    android:layout_toRightOf="@id/unright"
    android:background="@drawable/eight"/>"
</RelativeLayout>
```

2.2.3 TableLayout（表格布局）

表格布局中每一行为一个TableRow对象，在TableRow对象中可以再添加子控件，每添加一个即为一列。

表格布局中有几个重要的属性，其说明如表2-6所示。

表 2-6 表格布局的属性

xml属性	作用
android:stretchColumns	指定拉伸列（从0开始计数）。当所有列的内容不能填满整个TableLayout时，会拉伸指定的列，使其宽度变宽，达到填满整个父控件的目的
android:layout_column	该控件在TableRow中所处的列
android:layout_span	该控件所跨越的列数
android:collapseColumns	将指定的列隐藏，若有多列需要隐藏，用逗号将列序号隔开

下面通过一个简单的案例来学习表格布局以及设置控件的属性。

【案例2-3】：用TableLayout编写出如图2-3所示的界面。

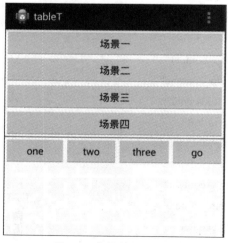

图 2-3 表格布局效果图

1. 案例分析

本案例采用表格布局实现。

（1）界面中所要显示的字符在"res\values\strings.xml"文件中定义。

（2）在"res\layout\activity_main.xml"文件中布局：添加5个TableRow对象，前面4个TableRow对象中分别放1个Button控件，后面1个TableRow对象中放入4个Button控件。

2. 操作步骤

步骤01 创建一个Android工程，命名为"TableLayout"。

步骤02 展开"Package Explorer"窗口中的"TableLayout"项目，打开"res\values\strings.xml"文件，修改并输入一些代码，代码如下。

```xml
<?xml version="1.0" encoding="utf-8"?>
<resources>
    <string name="app_name">tableT</string>
    <string name="action_settings">Settings</string>
    <string name="hello_world">Hello world!</string>
    <string name="changjiyi">场景一</string>
    <string name="changjier">场景二</string>
    <string name="changjisan">场景三</string>
    <string name="changjisi">场景四</string>
    <string name="one">one</string>
    <string name="two">two</string>
    <string name="three">three</string>
    <string name="go">go</string>
</resources>
```

> **提示**：一般情况下，在项目中要显示的文本是在"strings.xml"文件中设置的。

步骤03 打开"res\layout\activity_main.xml"文件，修改并输入一些代码，代码如下。

```xml
<?xml version="1.0" encoding="utf-8"?>
<TableLayout xmlns:android="http://schemas.android.com/apk/res/android"
    android:layout_width="fill_parent"
    android:layout_height="fill_parent"
    >
    <TableRow>
        <Button
            android:id="@+id/b1"
            android:layout_width="fill_parent"
```

```xml
            android:layout_height="wrap_content"
            android:layout_weight="1"
            android:text="@string/changjiyi" />
    </TableRow>
    <TableRow>
        <Button
            android:id="@+id/b2"
            android:layout_width="fill_parent"
            android:layout_height="wrap_content"
            android:layout_weight="1"
            android:text="@string/changjier" />
    </TableRow>
    <TableRow>
        <Button
            android:id="@+id/b3"
            android:layout_width="wrap_content"
            android:layout_height="wrap_content"
            android:layout_weight="1"
            android:text="@string/changjisan" />
    </TableRow>
    <TableRow>
        <Button
            android:id="@+id/b4"
            android:layout_width="wrap_content"
            android:layout_height="wrap_content"
            android:layout_weight="1"
            android:text="@string/changjisi" />
    </TableRow>
    <View
        android:layout_height="2dip"
        android:background="#FF909090" />
    <TableRow>
        <Button
            android:id="@+id/b5"
            android:layout_width="wrap_content"
            android:layout_height="wrap_content"
            android:layout_weight="1"
```

```
        android:text="@string/one" />
    <Button
        android:id="@+id/b6"
        android:layout_width="wrap_content"
        android:layout_height="wrap_content"
        android:layout_weight="1"
        android:text="@string/two" />
    <Button
        android:id="@+id/b7"
        android:layout_width="wrap_content"
        android:layout_height="wrap_content"
        android:layout_weight="1"
        android:text="@string/three" />
    <Button
        android:id="@+id/b8"
        android:layout_width="wrap_content"
        android:layout_height="wrap_content"
        android:layout_weight="1"
        android:text="@string/go" />
  </TableRow>
</TableLayout>
```

2.3 文本框及按钮控件

本节主要讲解3个常用控件：TextView（文本显示控件）、EditText（实现文本域，即可在此文本域中输入内容）和Button（按钮控件）。这些控件有一些常用的xml属性，如表2-7所示。

表 2-7 常用控件的xml属性

xml属性	作用
android:layout_width	指定控件的宽度
android:layout_height	指定控件的高度
android:id	为控件指定相应的ID
android:text	指定控件中显示的文字，尽量使用"strings.xml"文件
android:gravity	指定控件的基本位置
android:textSize	指定控件中字体的大小

（续表）

xml属性	作用
android:background	指定该控件所使用的背景颜色，使用RGB颜色模式，也可以引用android:drawable中的图片
android:padding	指定控件的内边距
android:singleLine	如果设置为true，则控件当中的内容在同一行显示，多余内容用省略号表示

下面通过一个简单的案例学习文本显示控件、实现文本域控件和按钮控件的使用。

【案例2-4】：运用文本显示控件、实现文本域控件和按钮控件编写邮箱登录界面，界面效果如图2-4所示。

图 2-4　邮箱登录界面效果图

1. 案例分析

本案例采用表格布局来实现。

（1）界面中要显示的字符在"res\values\strings.xml"文件中定义。

（2）在"res\layout\activity_main.xml"文件中进行布局，使用相对布局。添加3个文本显示控件、2个实现文本域控件和2个按钮控件。

2. 操作步骤

步骤01　创建一个Android工程，命名为"kongjian"。

步骤02　展开"Package Explorer"窗口中的"kongjian"项目，打开"res\valuest\strings.xml"文件，修改并输入一些代码，代码如下。

```
<?xml version="1.0" encoding="utf-8"?>
<resources>
  <string name="app_name">简易邮箱界面</string>
  <string name="action_settings">Settings</string>
  <string name="hello_world">Hello world!</string>
```

```xml
    <string name="mail">Mail</string>
    <string name="address">账号</string>
    <string name="pwd">密码</string>
    <string name="submit">确认</string>
    <string name="no">取消</string>
</resources>
```

步骤 03 打开"res\layout\activity_main.xml"文件，修改并输入一些代码，代码如下。

```xml
<?xml version="1.0" encoding="utf-8"?>
<RelativeLayout xmlns:android="http://schemas.android.com/apk/res/android"
    android:layout_width="wrap_content"
    android:layout_height="wrap_content"
    android:orientation="vertical" >

    <TextView
        android:id="@+id/tvt"
        android:layout_width="wrap_content"
        android:layout_height="wrap_content"
        android:layout_alignParentTop="true"
        android:text="@string/mail"
        android:textSize="80px" >
    </TextView>

    <Button
        android:id="@+id/btn_cancel"
        android:layout_width="wrap_content"
        android:layout_height="wrap_content"
        android:layout_alignBaseline="@+id/btn_login"
        android:layout_alignBottom="@+id/btn_login"
        android:layout_marginLeft="44dp"
        android:layout_toRightOf="@+id/btn_login"
        android:text="@string/no" />

    <TextView
        android:id="@+id/tv_password"
        android:layout_width="wrap_content"
        android:layout_height="wrap_content"
```

```xml
    android:layout_below="@+id/tv_username"
    android:layout_marginTop="48dp"
    android:text="@string/pwd"
    android:textSize="40px" />

<TextView
    android:id="@+id/tv_username"
    android:layout_width="wrap_content"
    android:layout_height="wrap_content"
    android:layout_below="@+id/tvt"
    android:layout_marginTop="24dp"
    android:text="@string/address"
    android:textSize="40px" />

<EditText
    android:id="@+id/txt_password"
    android:layout_width="fill_parent"
    android:layout_height="wrap_content"
    android:layout_alignParentLeft="true"
    android:layout_below="@+id/tv_username"
    android:ems="10" >
    <requestFocus />
</EditText>

<EditText
    android:id="@+id/EditText01"
    android:layout_width="fill_parent"
    android:layout_height="wrap_content"
    android:layout_alignParentLeft="true"
    android:layout_below="@+id/tv_password"
    android:ems="10" />

<Button
    android:id="@+id/btn_login"
    android:layout_width="wrap_content"
    android:layout_height="wrap_content"
    android:layout_below="@+id/EditText01"
    android:layout_marginLeft="30dp"
```

```
android:layout_marginTop="44dp"
android:layout_toRightOf="@+id/tv_password"
android:text="@string/submit" />
```

</RelativeLayout>

课后作业

一、填空题

1. View（视图）是Android中最基本的一个类，_____和_____都是继承自View类。

2. Android开发中一般分为以下几种布局：_____、_____、_____、_____和_____。

3. 布局主要是在"_____"文件中编写。

4. 相对布局容器内子组件的位置总是相对_____、_____来决定的，因此这种布局方式被称为相对布局。

二、选择题

1. 相对布局中设置控件与给定控件关系的属性中，将该控件的底部边缘与给定ID控件的底部边缘对齐的属性是（   ）。

    A. android:layout_alignBaseline

    B. android:layout_toRightOf

    C. android:layout_alignBottom

    D. android:layout_toLeftOf

2. 以下为表格布局的局部属性的是（   ）。

    A. android:collapseColumns

    B. android:layout_column

    C. android:stretchColumns

    D. android:shrinkColumns

3. Android不提供任何布局控制，而是开发人员自己通过X、Y两个坐标来控制组件的位置。当使用绝对布局作为布局容器时，布局容器不再管理组件的位置、大小——这些都需要开发人员自己控制的布局是（   ）。

    A. TableLayout

    B. ConstraintLayout

    C. RelativeLayout

    D. AbsoluteLayout

4. 在相对布局中设置控件方向的属性中，将该控件在其父控件范围内垂直且水平居中的属性是（　　）。

　　A. android:layout_centerHorizontal

　　B. android:layout_centerInparent

　　C. android:orientation

　　D. android:layout_centerVertical

三、操作题

参照书中所讲内容，练习使用Android Studio创建项目，分别使用线性布局和相对布局进行布局设置。通过构建同一个用户界面，加深对每种布局特点的理解；根据布局文件的复杂程度，分析这些布局的优缺点。这样在实际操作中学习，可以更好地理解布局方式选择的重要性。

第 3 章

菜单和对话框

内容概要

菜单是用户界面中最常见的元素之一，使用非常频繁。在手机应用程序中，由于受到手机屏幕大小的制约，菜单在手机应用中的使用较少。当图形用户界面在前台运行时，用户若按下手机上的Menu键，就会在屏幕底端弹出相应的选项菜单，其对应功能需要程序开发者编程实现。在Android系统中，菜单分为3种，分别是选项菜单（OptionsMenu）、上下文菜单（ContextMenu）和子菜单（SubMenu）。本章将介绍在Android中菜单和对话框的创建和使用。

数字资源

【本章案例文件】："案例文件\第3章"目录下

- 配套资源
- 入门精讲
- 项目实战
- 日志记录

3.1 选项菜单和子菜单

在Android中,由于屏幕等资源的限制,一个菜单中最多只会显示数量较少的菜单项,超出部分会自动隐藏,而且会出现一个"更多"的菜单项提示用户。在Android开发中,基于相关资源的限制,如何使用好菜单控件是Android程序开发人员要仔细思考的。

Android开发中菜单的创建有两种方式,第一种是动态创建,也就是通过在Activity中重载onCreateOptionMenu方法来创建菜单,一般使用menu.add("选项名")方法添加一个菜单选项,使用menu.getItem(int itemIndex)方法根据选项的索引获取该菜单选项,并对其进行操作,例如,添加点击事件onMenuItemClickListener,用addSubMenu方法添加子菜单等。这种方法的优点是创建方便,不需要建立资源文件,而缺点是对于菜单条目过多或多级菜单操作不方便,按键多的时候添加点击事件比较麻烦。

另一种方法是通过XML资源文件创建,也就是在res文件夹下创建名为menu的文件夹,再通过该文件夹下的"menu.xml"文件创建菜单样式,然后再利用getMenuInflater().inflate(menuRes, menu)方法将创建的菜单样式inflate到程序中。这种方法大体分为三步:首先,建立XML资源文件;其次,在onCreateOptionMenu回调函数完成XML资源的解析(用inflate方法),把菜单资源文件压入到程序中,不同菜单使用不同的回调函数,如ContextMenu菜单使用onCreateContextMenu函数;最后,如果需要给菜单添加点击事件,一般是通过重载onMenuItemSelected方法来设置,不同菜单使用不同的点击回调函数,如ContextMenu菜单使用onContextItemSelected函数。

■3.1.1 创建OptionsMenu菜单实例

选项菜单OptionsMenu的默认样式是在屏幕底部弹出一个菜单,创建方式有两种:第一种是通过Menu类创建菜单,第二种是通过XML格式的布局文件添加菜单的样式。下面通过案例来展示如何使用这两种方式创建选项菜单。

【案例3-1】:使用OptionsMenu设计一个界面,效果如图3-1所示。当选择某个选项时,会显示一个提示信息。

图 3-1 界面效果图

方法一:通过Menu类创建菜单

重载onCreateOptionsMenu(Menu menu)方法,并在此方法中通过Menu类的add方法添加菜单项,add方法中有4个参数,依次代表组别、ID、显示顺序和菜单的显示文本,操作步骤如下。

步骤 01 创建项目"MenuDemo",并创建一个名为"DefaultMenu"的Activity。

步骤 02 重载onCreateOptionsMenu(Menu menu)方法，并在此方法中添加菜单项，最后返回true；如果返回false，则不会显示菜单。

步骤 03 使用onOptionsItemSelected(MenuItem item)方法为菜单项注册事件。

步骤 04 其他按需要重载。例如，重载onOptionsMenuClosed(Menu menu)方法可以处理菜单关闭后发生的动作等。

完整代码如下。

```
package com.chapt5;

import android.app.Activity;
import android.os.Bundle;
import android.view.Menu;
import android.view.MenuItem;
import android.widget.Toast;

public class DefaultMenu extends Activity {
    /** Called when the activity is first created. */
    public void onCreate(Bundle savedInstanceState) {
        super.onCreate(savedInstanceState);
        setContentView(R.layout.main);
    }
    public boolean onCreateOptionsMenu(Menu menu) {
        /* * * add方法的4个参数依次是:
         * 1.组别，如果不分组的话就写Menu.NONE* *
         * 2.ID，这个很重要，Android根据这个ID来确定不同的菜单 *
         * 3.顺序，哪个菜单在前面由这个参数的大小决定 * *
         * 4.文本，菜单的显示文本 */
        menu.add(Menu.NONE, Menu.FIRST + 1, 5, "删除").setIcon(
            android.R.drawable.ic_menu_delete);
        // 可用setIcon方法为菜单设置图标，这里使用的是系统自带的图标
        menu.add(Menu.NONE, Menu.FIRST + 2, 2, "保存").setIcon(
            android.R.drawable.ic_menu_save);
        menu.add(Menu.NONE, Menu.FIRST + 3, 6, "帮助").setIcon(
            android.R.drawable.ic_menu_help);
        menu.add(Menu.NONE, Menu.FIRST + 4, 1, "添加").setIcon(
            android.R.drawable.ic_menu_add);
        menu.add(Menu.NONE, Menu.FIRST + 5, 4, "详细").setIcon(
            android.R.drawable.ic_menu_info_details);
```

```java
        menu.add(Menu.NONE, Menu.FIRST + 6, 3, "发送").setIcon(
            android.R.drawable.ic_menu_send);
        return true;
    }

    public boolean onOptionsItemSelected(MenuItem item) {
        switch (item.getItemId()) {
        case Menu.FIRST + 1:
            Toast.makeText(this, "删除菜单被点击了", Toast.LENGTH_LONG).show();
            break;
        case Menu.FIRST + 2:
            Toast.makeText(this, "保存菜单被点击了", Toast.LENGTH_LONG).show();
            break;
        case Menu.FIRST + 3:
            Toast.makeText(this, "帮助菜单被点击了", Toast.LENGTH_LONG).show();
            break;
        case Menu.FIRST + 4:
            Toast.makeText(this, "添加菜单被点击了", Toast.LENGTH_LONG).show();
            break;
        case Menu.FIRST + 5:
            Toast.makeText(this, "详细菜单被点击了", Toast.LENGTH_LONG).show();
            break;
        case Menu.FIRST + 6:
            Toast.makeText(this, "发送菜单被点击了", Toast.LENGTH_LONG).show();
            break;
        }
        return false;
    }

    public void onOptionsMenuClosed(Menu menu) {
        Toast.makeText(this, "选项菜单关闭了", Toast.LENGTH_LONG).show();
    }
}
```

方法二：通过XML布局文件实现菜单

通过布局Layout创建菜单对应的XML文件，然后在onCreateOptionsMenu方法中设置menu为已定义的"res\menu\menu.xml"文件，操作步骤如下。

步骤01 新建项目"MenuDemo"。

步骤 02 在"strings.xml"文件中输入需要显示的字符。打开"res\values\strings.xml"文件，代码如下。

```xml
<?xml version="1.0" encoding="utf-8"?>
<resources>
    <string name="hello">Hello World, DefaultMenu!</string>
    <string name="app_name">menudemo</string>
    <string name="add">添加</string>
    <string name="save">保存</string>
    <string name="send">发送</string>
    <string name="detail">详细</string>
    <string name="delete">删除</string>
    <string name="help">帮助</string>
</resources>
```

步骤 03 执行"新建"→"Other"→"Android"→"Android XML File"命令，创建名为"menu.xml"文件，该文件内容如下。

```xml
<?xml version="1.0" encoding="utf-8"?>
<menu xmlns:android="http://schemas.android.com/apk/res/android">
    <item android:id="@+id/add" android:title="@string/add"
        android:icon="@android:drawable/ic_menu_add" />
    <item android:id="@+id/save" android:title="@string/save"
        android:icon="@android:drawable/ic_menu_save" />
    <item android:id="@+id/send" android:title="@string/send"
        android:icon="@android:drawable/ic_menu_send" />
    <item android:id="@+id/detail" android:title="@string/detail"
        android:icon="@android:drawable/ic_menu_info_details" />
    <item android:id="@+id/delete" android:title="@string/delete"
        android:icon="@android:drawable/ic_menu_delete" />
    <item android:id="@+id/help" android:title="@string/help"
        android:icon="@android:drawable/ic_menu_help" />
</menu>
```

步骤 04 创建一个名为"DefaultMenu"的Activity，重载onCreateOptionsMenu(Menu menu)方法，并使用onOptionsItemSelected(MenuItem item)方法为菜单项注册事件。

```
package com.chapt5;
import android.app.Activity;
import android.os.Bundle;
```

```java
import android.view.Menu;
import android.view.MenuInflater;
import android.view.MenuItem;
import android.widget.Toast;
public class DefaultMenu extends Activity {
    public void onCreate(Bundle savedInstanceState) {
        super.onCreate(savedInstanceState);
        setContentView(R.layout.main);
    }
    public boolean onCreateOptionsMenu(Menu menu) {
        MenuInflater inflater = getMenuInflater();
        inflater.inflate(R.menu.menu, menu);
        return true;
    }
    public boolean onOptionsItemSelected(MenuItem item) {
        switch (item.getItemId()) {
        case R.id.delete:
            Toast.makeText(this, "删除菜单被点击了", Toast.LENGTH_LONG).show();
            break;
        case R.id.save:
            Toast.makeText(this, "保存菜单被点击了", Toast.LENGTH_LONG).show();
            break;
        case R.id.help:
            Toast.makeText(this, "帮助菜单被点击了", Toast.LENGTH_LONG).show();
            break;
        case R.id.add:
            Toast.makeText(this, "添加菜单被点击了", Toast.LENGTH_LONG).show();
            break;
        case R.id.detail:
            Toast.makeText(this, "详细菜单被点击了", Toast.LENGTH_LONG).show();
            break;
        case R.id.send:
            Toast.makeText(this, "发送菜单被点击了", Toast.LENGTH_LONG).show();
            break;
        }
        return false;
    }
    public void onOptionsMenuClosed(Menu menu) {
```

```
Toast.makeText(this, "选项菜单关闭了", Toast.LENGTH_LONG).show();
  }
}
```

> **提示**：(1) 用XML布局文件实现时，在onCreateOptionsMenu（Menu menu）方法中，需要通过MenuInflater对象把"menu.xml"文件转化为menu对象，而通过menu.add方法则是直接添加菜单项来构建菜单。
> (2) Toast是Android中用来显示信息的一种机制，和Dialog不一样的是：Toast没有焦点，而且Toast显示的时间有限，在显示一定时间后就会自动消失。

■3.1.2 监听菜单事件

除了重载onOptionsItemSelected(MenuItem item)方法以处理菜单的点击事件外，还可以通过菜单项的setOnMenuItemClickListener方法为不同的菜单项分别绑定监听器。采用后者这种方式不需要为每个菜单项指定ID，而是通过获取所添加的MenuItem对象，然后为对象绑定监听者。以下给出"delete"菜单项的创建、事件注册与响应的代码。由于篇幅有限，其他菜单项的处理与其类似，在此不再赘述。

```
public boolean onCreateOptionsMenu(Menu menu) {
  MenuItem deleteItem = menu.add(Menu.NONE, Menu.FIRST + 1, 5, "删除").setIcon(android.R.drawable.ic_menu_delete);
  deleteItem.setOnMenuItemClickListener(new OnMenuItemClickListener() {
  public boolean onMenuItemClick(MenuItem arg0) {
  Toast.makeText(MenuDemo2.this, "删除菜单被点击了", Toast.LENGTH_LONG).show();
  return false;
  }
  });
  return true;
}
```

> **提示**：(1) 该方法实现效果与前一个完全相同，区别仅在于处理菜单事件的监听方式不同。一般来说，通过重载onOptionsItemSelected(MenuItem item)方法处理菜单的点击事件更加简洁，因为所有的事件处理代码都控制在该方法内；通过绑定事件监听器可使程序具有更清晰的逻辑，但缺点是代码显得有些臃肿。
> (2) 如果是通过XML布局文件实现的菜单，可以通过MenuItem delete=(MenuItem)findViewById(R.id.delete)语句获取菜单项对象。
> (3) 如果希望所创建的菜单项是单选菜单项或多选菜单项，则可以通过调用菜单项的setCheckable(Boolean checkable)方法来设置该菜单项是否可以被勾选，也可通过调用setGroupCheckable方法来设置组里的菜单是否可以被勾选。
> (4) 通过菜单项的setShortcut方法可以为其设置快捷键。

3.1.3 与菜单项关联的Activity的设置

在应用程序中，如果需要点击某个菜单项以启动其他Activity或Service时，不需要开发者编写任何事件处理代码，只要调用MenuItem的setIntent(Intent intent)方法即可。该方法把菜单项与指定的Intent对象关联在一起，当用户点击该菜单项时，该Intent对象所代表的组件将会启动。

【案例3-2】：单击"Start the other Activity"选项，启动另一个Activity，效果如图3-2所示。

图 3-2　效果图

操作步骤如下。

步骤01 创建项目"MenuitemActivity"。

步骤02 创建一个名为"MenuitemActivity"的Activity，并重载onCreate和onCreateOptionsMenu方法。

```
package com.chapt5;
import android.app.Activity;
import android.content.Intent;
import android.os.Bundle;
import android.view.Menu;
import android.view.MenuItem;
public class MenuitemActivity extends Activity {
  public void onCreate(Bundle savedInstanceState) {
    super.onCreate(savedInstanceState);
    setContentView(R.layout.main);
  }
  public boolean onCreateOptionsMenu(Menu menu) {
    MenuItem mi = menu.add("Start the other Activity");
```

```
        mi.setIntent(new Intent(this, OtherActivity.class));
        return super.onCreateOptionsMenu(menu);
    }
}
```

步骤03 创建另一个名为"OtherActivity"的Activity,然后在AndroidManifest.xml中注册此Activity,在其中添加如下代码。

```
<activity android:name=".OtherActivity"
    android:label="This is Other Activity!" />
```

3.2 上下文菜单

Android中ContextMenu代表上下文菜单,类似于桌面程序中单击右键弹出的菜单。在Android中,上下文菜单不是通过单击鼠标右键得到的,而是通过长时间按住界面上的元素出现的。开发上下文菜单的方法与选项菜单的方法基本相似,因为ContextMenu也是Menu的子类,所以可用相同的方法为它添加菜单项。与开发选项菜单的区别在于:开发上下文菜单不是重载onCreateOptionsMenu(Menu menu)方法,而是调用onCreateContextMenu(ContextMenu menu,View source,ContextMenu.ContextMenuInfo menuInfo)方法。该方法在每次启动上下文菜单时都会被调用一次,该方法可以通过add方法添加相应的菜单项。

开发上下文菜单的过程如下所述。

(1) 重载onCreateContextMenu()方法。

(2) 调用Activity的registerForContextMenu(View view)方法为view组件注册上下文菜单。

(3) 重载onContextItemSelected(MenuItem mi)方法或者绑定事件监听器,对菜单项进行事件响应。

【案例3-3】:定义上下文菜单,先选择颜色,然后根据所选择的颜色更改文本框的背景颜色,效果如图3-3所示。

图 3-3 效果图

操作步骤如下。

步骤01 创建项目"ContextMenu"。

步骤02 创建一个名为"ContextMenuActivity"的Activity,并在onCreate方法中为文本框注册上下文菜单。

步骤03 重载onCreateContextMenu方法,在该方法中创建含有"红色""绿色""蓝色"和"退出"4个菜单项的菜单。

步骤04 重载onContextItemSelected方法对事件进行注册。

完整代码如下。

```java
package com.chapt5;
import android.app.Activity;
import android.graphics.Color;
import android.os.Bundle;
import android.view.ContextMenu;
import android.view.ContextMenu.ContextMenuInfo;
import android.view.Menu;
import android.view.MenuItem;
import android.view.View;
import android.widget.TextView;
public class ContextMenuActivity extends Activity {
    TextView text;
    public void onCreate(Bundle savedInstanceState) {
        super.onCreate(savedInstanceState);
        setContentView(R.layout.main);
        text = (TextView) findViewById(R.id.contextmenu);
        registerForContextMenu(text); // 为文本框注册上下文菜单
    }
    // 每次创建上下文菜单时都会触发该方法
    public void onCreateContextMenu(ContextMenu menu, View v,
            ContextMenuInfo menuInfo) {
        text.setText("这是关于上下文菜单的测试操作！"); // 设置提示信息
        menu.add(1, Menu.FIRST + 1, 1, "红色");
        menu.add(1, Menu.FIRST + 2, 2, "绿色");
        menu.add(1, Menu.FIRST + 3, 3, "蓝色");
        menu.add(1, Menu.FIRST + 4, 4, "退出");
        menu.setGroupCheckable(1, true, true);
        menu.setHeaderTitle("请选择：");
        menu.setHeaderIcon(android.R.drawable.ic_dialog_info);
    }
    public boolean onContextItemSelected(MenuItem item) {
        switch (item.getItemId()) {
        case Menu.FIRST + 1:
            text.setBackgroundColor(Color.RED);
            break;
        case Menu.FIRST + 2:
            text.setBackgroundColor(Color.GREEN);
```

```
            break;
        case Menu.FIRST + 3:
            text.setBackgroundColor(Color.BLUE);
            break;
        case Menu.FIRST + 4:
            ContextMenuActivity.this.finish();
            break;
        default:
            item.setChecked(true);
        }
        return true;
    }
}
```

3.3 Android中的对话框

Android中的对话框可以自定义，同时Android也对对话框提供了丰富的支持。常用的对话框有以下4种。

（1）AlertDialog：提示对话框，该对话框功能丰富、应用最广泛。

（2）ProgressDialog：进度对话框，该对话框只是对简单进度条的封装。

（3）DatePickerDialog：日期选择对话框，该对话框是对DatePicker的包装。

（4）TimePickerDialog：时间选择对话框，该对话框是对TimePicker的包装。

这4种对话框中功能最强、用法最灵活的就是AlertDialog，其应用得也最为广泛。

3.3.1 AlertDialog（提示对话框）

提示对话框是一个提示窗口，要求用户在此窗口内做出选择。该对话框中一般会有几个选择按钮、标题信息或提示信息。提示对话框提供了一些方法来生成4种预定义对话框。

（1）带消息、带n个按钮的提示对话框。

（2）带列表、带n个按钮的列表对话框。

（3）带多个单选列表项，带n个按钮的对话框。

（4）带多个多选列表项，带n个按钮的对话框。

提示对话框的构造方法全部是Protected类型的，所以不能直接通过提示对话框对象来创建对话框。要创建一个提示对话框，就要用到AlertDialog.Builder中的create方法。使用AlertDialog.Builder创建对话框需要了解以下几种方法。

（1）setTitle方法：为对话框设置标题。

（2）setIcon方法：为对话框设置图标。

（3）setMessage方法：为对话框设置内容。

（4）setView方法：为对话框设置自定义样式。

（5）setItems方法：用于设置对话框要显示的一个list，一般用于显示几个命令的列表。

（6）setMultiChoiceItems方法：用于设置对话框要显示的一系列复选框。

（7）setNeutralButton方法：为对话框设置普通按钮。

（8）setPositiveButton方法：为对话框添加"Yes"按钮。

（9）setNegativeButton方法：为对话框添加"No"按钮。

（10）create方法：用于创建对话框。

（11）show方法：用于显示对话框。

创建提示对话框的主要过程如下所述。

（1）获得提示对话框的静态内部类Builder对象，由该类创建对话框。

（2）通过Builder对象设置对话框的标题、按钮以及按钮将要响应的事件。

（3）调用Builder对象的create方法创建对话框。

（4）调用提示对话框的show方法显示对话框。

【案例3-4】：创建不同类型的对话框，所有的对话框定义在一个Activitiy中，本例只给出部分代码片段。

1. 创建内容输入对话框

```
new AlertDialog.Builder(this).setTitle("请输入").setIcon(android.R.drawable.ic_dialog_info)
  .setView( new EditText(this)).setPositiveButton("确定", null)
  .setNegativeButton("取消", null).show();
```

内容输入对话框的运行效果如图3-4所示。

图 3-4　内容输入对话框

2. 创建带按钮的提示对话框

```
Dialog alertDialog = new AlertDialog.Builder(this).setTitle("提示")
  .setMessage("您确定退出吗？"). setIcon(R.drawable.icon)
  .setPositiveButton("确定", new DialogInterface.OnClickListener() {
    public void onClick(DialogInterface dialog, int which) {
      dialog.dismiss();
      DialogDemoActivity.this.finish();
```

```
        }
    })
.setNegativeButton("取消", new DialogInterface.OnClickListener() {
    public void onClick(DialogInterface dialog, int which) {
        dialog.dismiss();
    }
}).create();
alertDialog.show();
```

带按钮的提示对话框的运行效果如图3-5所示。

图 3-5　带按钮的提示对话框

3. 创建列表对话框

用setItems(CharSequence[] items, final OnClickListener listener)方法实现类似ListView的AlertDialog。对话框的第1个参数是要显示数据的数组，第2个参数是点击某个item的触发事件。

```
final String[] arrayFruit = new String[] { "苹果","橘子","草莓","香蕉" };
Dialog alertDialog = new AlertDialog.Builder(this).setTitle("你喜欢吃哪种水果？").setIcon(R.drawable.icon)
.setItems(arrayFruit, new DialogInterface.OnClickListener() {
    public void onClick(DialogInterface dialog, int which) {
        Toast.makeText(DialogDemoActivity.this,
    arrayFruit[which], Toast.LENGTH_SHORT).show();
        }})
.setNegativeButton("取消", new DialogInterface.OnClickListener() {
    public void onClick(DialogInterface dialog, int which) {
    }
}).create();
alertDialog.show();
```

列表对话框的运行效果如图3-6所示。

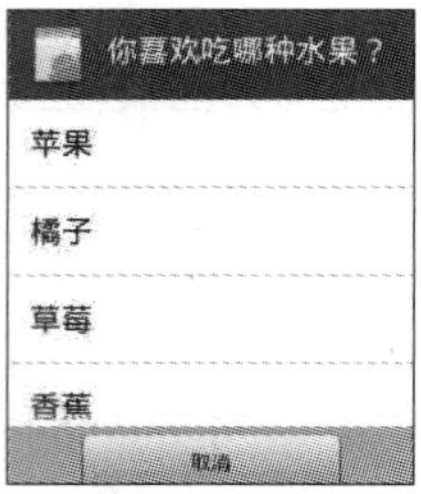

图 3-6 列表对话框

4. 创建单选列表对话框

用setSingleChoiceItems(CharSequence[] items, int checkedItem, final OnClickListener listener)方法实现类似RadioButton的AlertDialog。对话框的第1个参数是要显示数据的数组，第2个参数是初始值（初始被选中的item），第3个参数是点击某个item的触发事件。

```
final String[] arrayFruit = new String[] { "苹果","橘子","草莓","香蕉" };
Dialog alertDialog = new AlertDialog.Builder(this).setTitle("你喜欢吃哪种水果？ ").setIcon(R.drawable.icon)
    .setSingleChoiceItems(arrayFruit, 0,
      new DialogInterface.OnClickListener() {
        public void onClick(DialogInterface dialog, int which) {
          selectedFruitIndex = which;
        }
}).setPositiveButton("确认", new DialogInterface.OnClickListener() {
  public void onClick(DialogInterface dialog, int which) {
    Toast.makeText(DialogDemoActivity.this, arrayFruit[selectedFruitIndex],
Toast.LENGTH_SHORT).show();   }
      })
    .setNegativeButton("取消", new DialogInterface.OnClickListener() {
      public void onClick(DialogInterface dialog, int which) {
      }
      }).create();
  alertDialog.show();
```

单选列表对话框的运行效果如图3-7所示。

图 3-7 单选列表对话框

5. 创建多选列表对话框

用setMultiChoiceItems(CharSequence[] items, boolean[] checkedItems, final OnMultiChoiceClickListener listener)方法实现类似CheckBox的AlertDialog。对话框的第1个参数是要显示数据的数组，第2个参数是选中状态的数组，第3个参数是点击某个item的触发事件。

```
final String[] arrayFruit = new String[] { "苹果","橘子","草莓","香蕉" };
final boolean[] arrayFruitSelected = new boolean[] { true, true, false,false };
Dialog alertDialog = new AlertDialog.Builder(this).setTitle("你喜欢吃哪种水果？").setIcon(R.drawable.icon)
    .setMultiChoiceItems(arrayFruit, arrayFruitSelected,
    new DialogInterface.OnMultiChoiceClickListener() {
        public void onClick(DialogInterface dialog,int which,
            boolean isChecked) {
            arrayFruitSelected[which] = isChecked;
        }
}).setPositiveButton("确认", new DialogInterface.OnClickListener() {
    public void onClick(DialogInterface dialog, int which) {
        StringBuilder stringBuilder = new StringBuilder();
        for (int i = 0; i < arrayFruitSelected.length; i++) {
            if (arrayFruitSelected[i] == true) {
                stringBuilder.append(arrayFruit[i] + "、");
            }
        }
        Toast.makeText(DialogDemoActivity.this,
            stringBuilder.toString(), Toast.LENGTH_SHORT).show();
    }
}).setNegativeButton("取消", new DialogInterface.OnClickListener() {
    public void onClick(DialogInterface dialog, int which) {
    }
}).create();
alertDialog.show();
```

多选列表对话框的运行效果如图3-8所示。

图3-8 多选列表对话框

6. 创建自定义对话框

当系统自带的提示对话框的风格无法满足用户需求时,用户可以自己定义对话框。

【案例3-5】:创建用户登录对话框,对话框包含"用户名"和"密码"两个内容输入对话框及"登录"和"取消"两个按钮。自定义对话框的运行效果如图3-9所示。

图3-9 自定对话框

操作步骤如下。

步骤01 创建项目"DialogDemo"和一个名为"DialogDemoActivity"的Activity。

步骤02 创建登录界面的布局文件"login.xml",代码如下。

```
<?xml version="1.0" encoding="utf-8"?>
<LinearLayout xmlns:android="http://schemas.android.com/apk/res/android"
    android:layout_width="match_parent"
    android:layout_height="match_parent"
    android:orientation="vertical">
<LinearLayout
    android:layout_width="fill_parent"
    android:layout_height="wrap_content"
    android:gravity="center">
<TextView android:layout_width="0dip" android:layout_height="wrap_content"
    android:layout_weight="1"
    android:text="@string/username" />
<EditText android:layout_width="0dip" android:layout_height="wrap_content"
    android:layout_weight="1" />
    </LinearLayout>
<LinearLayout
    android:layout_width="fill_parent"
    android:layout_height="wrap_content"
    android:gravity="center">
<TextView android:layout_width="0dip" android:layout_height="wrap_content"
    android:layout_weight="1"
        android:text="@string/password" />
```

```
<EditText android:layout_width="0dip" android:layout_height="wrap_content"
    android:layout_weight="1" />
  </LinearLayout>
</LinearLayout>
```

步骤 03 重载"DialogDemoActivity"的onCreate方法，并添加如下代码。

```
LayoutInflater layoutInflater = LayoutInflater.from(this);
View myLoginView = layoutInflater.inflate(R.layout.login, null);
Dialog alertDialog = new AlertDialog.Builder(this).setTitle("用户登录")
    .setIcon(R.drawable.ic_launcher)
    .setView(myLoginView)
    .setPositiveButton("登录", new DialogInterface.OnClickListener() {
  public void onClick(DialogInterface dialog, int which) {
    }
}).setNegativeButton("取消", new DialogInterface.OnClickListener() {
  public void onClick(DialogInterface dialog, int which) {
    }
  }).create();
alertDialog.show();
```

3.3.2 ProgressDialog（进度对话框）

进度对话框类继承自提示对话框类，同样存放在android.app包中。进度对话框有两种形式：一种是圆圈旋转形式，一种是水平进度条形式。开发者可以通过属性设置修改其形式，也可以通过该类提供的一系列set方法设置对话框中进度条的风格、进度条的最大值等属性。

【案例3-6】：在主界面上放置一个命令按钮，点击命令按钮时会弹出一个进度对话框，提示后台正在执行程序。进度对话框的运行效果如图3-10所示。

图 3-10　进度对话框

操作步骤如下。

步骤01 创建项目"ProgressDialogDemo"和一个名为"ProgressDialogActivity"的Activity。

步骤02 修改"main.xml"文件，其中放置一个命令按钮，其ID设为button，其text设为"@string/execute"。

步骤03 修改"strings.xml"文件，代码如下。

```xml
<?xml version="1.0" encoding="utf-8"?>
<resources>
    <string name="app_name">ProgressDialogDemo</string>
    <string name="execute">执行</string>
    <string name="str_dialog_title">请稍等片刻</string>
    <string name="str_dialog_body">正在执行...</string>
</resources>
```

步骤04 修改"ProgressDialogActivity"类文件，代码如下。

```java
package com.chapt5;
import android.app.Activity;
import android.app.ProgressDialog;
import android.os.Bundle;
import android.view.View;
import android.view.View.OnClickListener;
import android.widget.Button;
public class ProgressDialogActivity extends Activity {
    private Button button=null;
    public ProgressDialog dialog=null;
    public void onCreate(Bundle savedInstanceState) {
        super.onCreate(savedInstanceState);
        setContentView(R.layout.main);
        button=(Button)findViewById(R.id.button);
        button.setOnClickListener(new OnClickListener(){
            public void onClick(View v) {
                String title =ProgressDialogActivity.this.getString(R.string.str_dialog_title);
                String body =ProgressDialogActivity.this.getString(R.string.str_dialog_body);
                //显示Progress对话框
                dialog=ProgressDialog.show(ProgressDialogActivity.this,strDialogTitle,strDialogBody,true);
                new Thread(){
                    public void run(){
                        try{
```

```
            //表示后台运行的代码段,以暂停3秒代替
            sleep(3000);
        }catch (InterruptedException e){
            e.printStackTrace();
        }finally{
            //卸载dialog对象
            dialog.dismiss();
        }
    }
    }.start();
    }
});
}
}
```

3.4 提示信息

在某些情况下需要向用户弹出提示消息,如显示错误信息、收到短消息等。Android提供有弹出消息、状态栏提醒等机制实现此功能。

3.4.1 Toast

Toast是Android中用来显示提示信息的一种机制。这个提示信息框用于向用户生成简单的提示信息。与对话框不同,Toast没有焦点,显示的时间有限,超过信息浮动显示设定的时长后会自动消失。创建Toast的主要过程如下所述。

(1) 调用Toast的构造器或者静态方法makeText创建一个Toast对象。
(2) 调用Toast的方法设置该消息提示的对齐方式、显示内容、显示时长等属性。
(3) 调用Toast的show方法将提示信息显示出来。

Toast一般用于显示简单的提示信息。显示较为复杂的信息,如图片、列表等,一般要用对话框完成。较为复杂的信息也可以用Toast的setView(view)方法以添加view组件的方式来实现,该方法允许用户自定义显示内容。创建Toast的常用方法如下。

```
Toast t = Toast.makeText(Context,msg,Toast.LENGTH_SHORT/LENGTH_LONG);
```

例如,在运行中弹出一个Toast,其提示信息为"你的愿望能实现"。

```
Toast.makeText(getApplicationContext(),"你的愿望能实现", Toast.LENGTH_SHORT).show()
```

3.4.2 Notification

Notification是Android提供的状态栏的提醒机制。手机的状态栏位于手机屏幕的最上方，那里一般显示手机当前的网络状态、电池状态、事件等。Notification不会打断用户当前的操作，支持异步的点击事件响应。程序一般由NotificationManager来管理，NotificationManager负责发通知、清除通知等，它是一个系统Service，必须通过getSystemService方法才能获取。创建Notification的步骤如下所述。

（1）通过getSystemService方法得到NotificationManager。
（2）构造一个Notification对象。
（3）设置Notification对象的属性参数。
（4）通过NotificationManager发送一个Notification。

在界面上放置一个命令按钮，点击命令按钮时创建一个Notification对象，其核心代码如下。

```
//得到NotificationManager
NotificationManager notificationManager = (NotificationManager) getSystemService(Context.NOTIFICATION_SERVICE);
//实例化一个Notification对象，并在实例化时设置图标、文本内容、发送时间
Notification notification = new Notification(R.drawable.image1, "notice", System.currentTimeMillis());
//创建一个启动其他Activity的intent对象
Intent intent = new Intent(ThisActivity.this,OtherActivity.class);
//创建一个延迟发送的Intent—PendingIntent对象
PendingIntent pendingIntent
= PendingIntent.getActivity(getApplicationContext(), 0, intent, 1);
//设置事件信息
notification.setLatestEventInfo(getApplicationContext(), "title","a message", pendingIntent);
//发送通知
notificationManager.notify(NOTIFICATION_ID, notification);
```

如果想要取消一个Notification，只需要使用NotificationManager的cancel方法即可实现，其核心代码如下。

```
NotificationManager notificationManager = (NotificationManager) getSystemService(Context.NOTIFICATION_SERVICE);
notificationManager.cancel(NOTIFICATION_ID);
```

课后作业

一、填空题

1. Android的SDK提供了3种类型的菜单，分别为_____，_____和_____。

2. Android开发中有两种创建菜单的方法，第1种方法是通过_____；第2种方法是通过_____。

3. _____继承了Menu，它代表了一个_____，实际就是将功能相同或相似的分组进行_____的一种菜单。

二、选择题

1. 常用的对话框有4种，（ ）是进度对话框。

    A. AlertDialog

    B. TimePickerDialog

    C. ProgressDialog

    D. DatePickerDialog

2. 下列不属于创建AlertDialog的主要步骤的是（ ）。

    A. 创建Andorid项目

    B. 获得AlertDialog的动态内部类Builder对象

    C. 调用Builder对象的create方法创建进度对话框

    D. 调用AlertDialog的show方法创建对话框

3. ProgressDialog类继承自（ ）类。

    A. Activity

    B. Service

    C. Menu

    D. AlertDialog

4. Android用（ ）代表上下文菜单，类似于桌面程序的右键弹出式菜单。

    A. optionsmenu_demo

    B. ContextMenu

    C. Menu

    D. addSubMenu

三、操作题

Android提供了丰富的对话框支持，其中功能最强、用法最灵活的就是AlertDialog。练习利用Android Studio开发环境创建一个新的Android应用程序，实现创建一个提示窗口的功能。（提示：参照书中3.3.1节内容完成。）

第4章

常用控件

内容概要

要设计出让用户喜欢的Android应用程序界面，除了要用到基本的TextView、EditText和Button控件外，还要用到更多其他的控件，如ImageButton控件、ImageView控件、RadioButton控件、CheckBox控件和ListView控件等。本章将详细介绍一些功能强大、应用广泛的控件。

数字资源

【本章案例文件】："案例文件\第4章"目录下

4.1 ImageButton控件

Android系统自带的除了Button按钮控件以外，还提供了带图标的ImageButton按钮控件。要制作带图标的按钮，首先要在布局文件中定义ImageButton，然后再通过以下两种方法（任选一种即可）设置要显示的图标。

方法一：在布局文件中直接设置按钮的图标，如android:src="@drawable/图片地址及图片名"。

ImageButton1.ImageDrawable(getResources().getDrawable(R.drawable.icon1));

方法二：使用系统自带的图标。具体操作如下所述。

设置好按钮的图标，然后再为按钮设置监听类setOnClickListener。下面通过一个简单的案例学习ImageButton控件及其属性设置。

【案例4-1】：使用ImageButton按钮设计一个界面，效果如图4-1所示。

图 4-1 ImageButton 案例效果图

1. 案例分析

首先，在"activity_main.xml"布局文件中添加一个TextView控件和ImageButton控件，并设置其属性。

其次，在"MainActivity.java"文件中定义一个变量，通过findViewById方法得到ImageButton控件，并添加对应的监听事件。

2. 操作步骤

步骤 01 创建一个Android工程，命名为"buttonimages"。

步骤 02 展开"Package Explorer"窗口中的"buttonimages"项目，然后打开"res\layout\activity_main.xml"文件，修改并输入一些代码，代码如下。

```xml
<LinearLayout xmlns:android="http://schemas.android.com/apk/res/android"
    xmlns:tools="http://schemas.android.com/tools"
    android:orientation="vertical" android:layout_width="fill_parent"
    android:layout_height="fill_parent"
    tools:context=".MainActivity" >
<TextView
    android:layout_width="fill_parent"
    android:layout_height="wrap_content"
    android:id="@+id/textView" />
<ImageButton
android:id="@+id/imageButton"
android:layout_width="wrap_content"
android:layout_height="wrap_content">
</ImageButton>
</LinearLayout>
```

步骤 03 打开 "src\com.example.buttonimages\MainActivity.java" 文件，修改并输入如下代码。

```java
package com.example.buttonimages;
import android.os.Bundle;
import android.app.Activity;
import android.view.View;
import android.widget.Button;
import android.widget.ImageButton;
import android.widget.TextView;
public class MainActivity extends Activity {
    @Override
    protected void onCreate(Bundle savedInstanceState) {
        super.onCreate(savedInstanceState);
        setContentView(R.layout.activity_main);
        setTitle("ImageButton");

        ImageButton imgButton = (ImageButton) this.findViewById(R.id.imageButton);
        // 设置图片按钮的背景
        imgButton.setBackgroundResource(R.drawable.buttonimage);
        // setOnClickListener()    响应图片按钮的点击事件监听器
```

```
imgButton.setOnClickListener(new Button.OnClickListener(){
@Override
public void onClick(View v) {
TextView txt = (TextView) MainActivity.this.findViewById(R.id.textView);
txt.setText(R.id.txtview);
}
});
}}
```

4.2 ImageView控件

ImageView控件是Android中基础图片的显示控件，也是布局中使用图片最常用的方式，可以让程序变得生动活泼。ImageView控件有个重要的属性是ScaleType，该属性用于表示图片显示的方式，共有8种取值。ScaleType属性值如表4-1所示。

表4-1 ScaleType属性值

ScaleType的值	描述
ScaleType.CENTER	图片大小为原始大小；如果图片大小大于ImageView控件的尺寸，则截取图片中间部分；若小于ImageView控件的尺寸，则直接将图片居中显示
ScaleType.CENTER_CROP	将图片等比例缩放，让图片的短边与ImageView边的长度相同，即不能留有空白，缩放后截取中间部分进行显示
ScaleType.CENTER_INSIDE	将尺寸大于ImageView的尺寸的图片进行等比例缩小，直到整幅图能够居中显示在ImageView中；尺寸小于ImageView尺寸的图片不变，直接居中显示
ScaleType.FIT_CENTER	ImageView的默认状态；大图等比例缩小，使整幅图能够居中显示在ImageView中；小图等比例放大，同样要整体居中显示在ImageView中
ScaleType.FIT_END	缩放方式同FIT_CENTER，只是将图片显示在右方或下方，而不是居中
ScaleType.FIT_START	缩放方式同FIT_CENTER，只是将图片显示在左方或上方，而不是居中
ScaleType.FIT_XY	将图片非等比例缩放到大小与ImageView相同
ScaleType.MATRIX	根据一个3×3矩阵对图片进行缩放

下面通过一个简单的案例来学习ImageView控件及其属性设置。

【案例4-2】：使用ImageView设计一个界面，效果如图4-2所示。

图 4-2　ImageView 案例效果图

操作步骤如下。

步骤 01 把图片导入到资源文件夹中，即将图片拖至项目中以"res\drawable"开头的5个文件夹下，它们分别是放置不同分辨率图片的文件夹。Android读取图片时会自动优化，会自动选用分辨率合适的图片显示。例如，高分辨率文件夹中可以存放128×128的图片，低分辨率文件夹中可以存放32×32的图片。

步骤 02 在"strings.xml"文件中输入需要显示的字符。打开"res\values\strings.xml"文件，修改并添加如下代码。

```
<?xml version="1.0" encoding="utf-8"?>
<resources>
    <string name="app_name">imageV</string>
    <string name="action_settings">Settings</string>
    <string name="hello_world">hello！</string>
</resources>
```

步骤 03 在XML布局文件中添加ImageView控件。打开"res\layout\activity_main.xml"文件，修改并添加如下代码。

```
<RelativeLayout xmlns:android="http://schemas.android.com/apk/res/android"
    xmlns:tools="http://schemas.android.com/tools"
    android:layout_width="match_parent"
```

```
        android:layout_height="match_parent"
        android:paddingBottom="@dimen/activity_vertical_margin"
        android:paddingLeft="@dimen/activity_horizontal_margin"
        android:paddingRight="@dimen/activity_horizontal_margin"
        android:paddingTop="@dimen/activity_vertical_margin"
        tools:context=".MainActivity" >
        <TextView
            android:layout_width="wrap_content"
            android:layout_height="wrap_content"
            android:text="@string/hello_world"
            android:textSize="80px" />
        <ImageView
            android:src="@drawable/one"
            android:layout_width="wrap_content"
            android:layout_height="wrap_content"></ImageView>
    </RelativeLayout>
```

4.3 单选按钮和复选框

单选按钮（RadioButton）和复选框（CheckBox）继承自Button类，因此可以直接使用Button控件支持的各种属性和方法。单选按钮和复选框与普通按钮的不同之处是多了一个可选中的功能，此功能需要一个额外的属性——android:checked属性，该属性用于指定选项是否被选中。

4.3.1 单选按钮组和单选按钮的用法

每一个单选按钮组里面至少包含两个单选按钮或更多个单选按钮，但只有一个单选按钮能被选中；不同组之间的单选按钮互不影响；每一个单选按钮组中都有一个默认被选中的单选按钮，大部分情况下建议选择第1个选项为默认选项。

【案例4-3】：使用单选按钮和单选按钮组设计一个界面，当选中某个单选项时，弹出相关的一段话。例如，当选中"海陆大餐（好吃真好吃）"选项时，则弹出「山珍海味」，乐不思蜀的人，为人海派，从不拖泥带水，拥有坚韧不拔的性格。但是不够冷静、过度挥霍，只怕会坐吃山空，不得不多加警惕"等语句。单选按钮效果如图4-3所示。

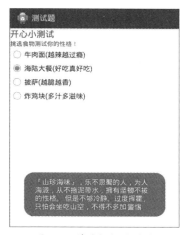

图 4-3　单选按钮效果图

操作步骤如下。

步骤 01 新建一个Android应用程序。

步骤 02 编写"strings.xml"文件,添加需要显示的字符。打开"res\values\strings.xml"文件,修改并添加一些代码,代码如下。

```xml
<?xml version="1.0" encoding="utf-8"?>
<resources>
    <string name="app_name">测试题</string>
    <string name="action_settings">Settings</string>
    <string name="title">开心小测试</string>
    <string name="choose">挑选食物测试你的性格!</string>
    <string name="niu">牛肉面(越辣越过瘾)</string>
    <string name="hai">海陆大餐(好吃真好吃)</string>
    <string name="pizza">披萨(越脆越香)</string>
    <string name="zha">炸鸡块(多汁多滋味)</string>
</resources>
```

步骤 03 编写"activity_main.xml"文件,添加一个RadioGroup标签,在RadioGroup标签内添加4个RadioButton。打开"res\layout\activity_main.xml"文件,修改并添加如下代码。

```xml
<LinearLayout xmlns:android="http://schemas.android.com/apk/res/android"
    xmlns:tools="http://schemas.android.com/tools"
    android:layout_width="match_parent"
    android:layout_height="match_parent"
    android:orientation="vertical"
    >
    <TextView
        android:layout_width="wrap_content"
        android:layout_height="wrap_content"
        android:text="@string/title"
        android:textSize="40px"
        />
    <TextView
        android:id="@+id/who"
        android:layout_width="wrap_content"
        android:layout_height="wrap_content"
        android:text="@string/choose"
        />
    <RadioGroup
```

```xml
    android:id="@+id/ceshi_group"
        android:layout_width="wrap_content"
        android:layout_height="wrap_content"
        android:orientation="vertical"
    >
    <RadioButton
        android:id="@+id/niunan"
        android:layout_height="wrap_content"
        android:layout_width="wrap_content"
        android:text="@string/niu"
        />
    <RadioButton
        android:id="@+id/hailu"
        android:layout_width="wrap_content"
        android:layout_height="wrap_content"
        android:text="@string/hai"
        android:checked="true"
        />
    <RadioButton
        android:id="@+id/pizza"
        android:layout_width="wrap_content"
        android:layout_height="wrap_content"
        android:text="@string/pizza"
        />
    <RadioButton
        android:id="@+id/zhaji"
        android:layout_width="wrap_content"
        android:layout_height="wrap_content"
        android:text="@string/zha"
        />
    </RadioGroup>
</LinearLayout>
```

步骤 04 编写Activity主文件：先声明6个全局变量，用于接收6个控件对象；在onCreate方法内，根据控件的ID获得这6个对象并赋给相应的变量；编写监听器。打开"src\com.example.sumothers\MainActivity.java"文件，修改并添加如下代码。

```java
package com.example.sumothers;
import android.app.Activity;
import android.os.Bundle;
import android.widget.CheckBox;
import android.widget.CompoundButton;
import android.widget.CompoundButton.OnCheckedChangeListener;
import android.widget.RadioButton;
import android.widget.RadioGroup;
import android.widget.TextView;
import android.widget.Toast;

public class MainActivity extends Activity {
    //定义各控件的变量
    private TextView who = null;
    private TextView how = null;
    private RadioGroup ceshi_group = null;
    private RadioButton niunan = null;
    private RadioButton hailu = null;
    private RadioButton pizza = null;
    private RadioButton zhaji = null;

    @Override
    public void onCreate(Bundle savedInstanceState) {
        super.onCreate(savedInstanceState);
        setContentView(R.layout.activity_main);

        //获得对应的控件
        who = (TextView)findViewById(R.id.who);
        ceshi_group = (RadioGroup)findViewById(R.id.ceshi_group);
        niunan = (RadioButton)findViewById(R.id.niunan);
        hailu = (RadioButton)findViewById(R.id.hailu);
        pizza = (RadioButton)findViewById(R.id.pizza);
        zhaji = (RadioButton)findViewById(R.id.zhaji);

        //设置ceshi_group的监听器，其实是一句代码，其参数是一个带有重构函数的对象
        ceshi_group.setOnCheckedChangeListener(new RadioGroup.OnCheckedChangeListener() {
            public void onCheckedChanged(RadioGroup group, int checkedId) {
```

```
        // TODO Auto-generated method stub
        if(checkedId == niunan.getId()){
            Toast.makeText(MainActivity.this,"吃辛辣食物的人，本身也很「辣」，性情孤傲，愤世嫉俗，对社交活动、对礼尚往来极端排斥，梦想成为立大功、成大业、永垂青史的英雄。", Toast.LENGTH_LONG).show();
        }
        else if(checkedId == hailu.getId()){
            Toast.makeText(MainActivity.this, "「山珍海味」，乐不思蜀的人，为人海派，从不拖泥带水，拥有坚韧不拔的性格。但是不够冷静、过度挥霍，只怕会坐吃山空，不得不多加警惕", Toast.LENGTH_LONG).show();
        }
        else if(checkedId == pizza.getId()){
            Toast.makeText(MainActivity.this, "喜欢吃「薄饼」的人，为人也比较刻薄小气，在团体中属于叛逆的角色，有点自以为是。", Toast.LENGTH_LONG).show();
        }
        else if(checkedId == zhaji.getId()){
            Toast.makeText(MainActivity.this, "这种人属于不爱动的后现代主义者，感情「脆」弱、深怕寂寞，举手投足像只小绵羊一般温驯，欠缺冲劲。", Toast.LENGTH_LONG).show();
        }
    }
});
    }
}
```

步骤05 运行程序，即可得到如图4-3所示的效果。

知识点拨

（1）监听器实现的是RadioGroup.OnCheckedChangeListener()提供的接口，需要重载里面的public void onCheckedChanged(RadioGroup group, int checkedId) 方法，此方法的第1个参数用于接收RadioGroup对象，第2个参数用于接收被选中的RadioButton的ID。在这个方法中可以做一系列的判断和操作，如判断RadioButton的ID是否等于checkedId，如果等于就使用Toast显示提示消息。

（2）Toast是Android中用于显示提示信息的一种机制，Toast没有焦点，而且Toast显示的时间有限，在显示一定时间后会自动消失。

> **提示**：将监听器绑定到RadioGroup的注意事项有以下两点。
> (1) 这里绑定监听器的是RadioGroup对象而不是RadioButton对象。
> (2) 这里的监听器实现的是RadioGroup.OnCheckedChangeListener()提供的接口。

4.3.2 复选框的用法

复选框（CheckBox）是一种双状态的按钮，可以选中或不选中，每次点击的时候可以选择是否被选中，复选框能同时选中多个。在用户界面中，复选框默认的是矩形的显示方式。复选框不同于单选按钮（RadioButton），每个选项都是一个复选框。对于事件监听，复选框与单选按钮是一样的，都是通过CompoundButton.OnCheckedChangeListener()监听的。Java文件中为每一个复选框都编写了一个监听器，该监听器实现的是CompoundButton.OnCheckedChangeListener()提供的接口，需要重载里面的public void onCheckedChanged(CompoundButton buttonView, boolean isChecked)方法，这个方法的第1个参数用于接收CompoundButton对象，第2个参数用于接收是否被选中。在这个方法里可以做一系列的判断和操作，如判断某个复选框有没有被选中。

【案例4-4】：使用单选按钮、单选按钮组和复选框设计一个界面，使选中单选按钮能显示选中的内容；选中多选按钮，也能显示选中的内容。显示效果如图4-4所示。

图 4-4 单选、多选按钮效果图

操作步骤如下。

步骤 01 编写"strings.xml"文件，添加需要显示的字符。打开"res\values\strings.xml"文件，修改并添加如下代码。

```
<?xml version="1.0" encoding="utf-8"?>
<resources>
    <string name="app_name">olympicGames</string>
    <string name="action_settings">Settings</string>
    <string name="hello_world">Hello world!</string>
```

```xml
<string name="who">Who will be the number one?</string>
<string name="china">中国</string>
<string name="america">美国</string>
<string name="others">其他</string>
<string name="how">How many golds medals will China win?</string>
<string name="less">30以下</string>
<string name="thirty">30~39</string>
<string name="forty">40~49</string>
<string name="fifty">50以上</string>
</resources>
```

步骤 02 编写"activity_main.xml"文件,添加一个单选按钮组的标签,在单选按钮组的标签内添加3个单选按钮、4个复选框和2个文本框,修改并添加如下代码。

```xml
<LinearLayout xmlns:android="http://schemas.android.com/apk/res/android"
    xmlns:tools="http://schemas.android.com/tools"
    android:layout_width="match_parent"
    android:layout_height="match_parent"
    android:orientation="vertical"
    >
    <TextView
        android:id="@+id/who"
        android:layout_width="wrap_content"
        android:layout_height="wrap_content"
        android:text="@string/who"
         />
    <RadioGroup
        android:id="@+id/who_group"
        android:layout_width="wrap_content"
        android:layout_height="wrap_content"
        android:orientation="vertical"
        >
        <RadioButton
            android:id="@+id/china"
            android:layout_height="wrap_content"
            android:layout_width="wrap_content"
            android:text="@string/china"
            android:checked="true"
             />
```

```xml
<RadioButton
    android:id="@+id/america"
    android:layout_width="wrap_content"
    android:layout_height="wrap_content"
    android:text="@string/america"
    />
<RadioButton
    android:id="@+id/others"
    android:layout_width="wrap_content"
    android:layout_height="wrap_content"
    android:text="@string/others"
    />
</RadioGroup>
<TextView
    android:id="@+id/how"
    android:layout_width="wrap_content"
    android:layout_height="wrap_content"
    android:text="@string/how"
    />
<CheckBox
    android:id="@+id/less"
    android:layout_width="wrap_content"
    android:layout_height="wrap_content"
    android:text="@string/less"
    />
<CheckBox
    android:id="@+id/thirty"
    android:layout_width="wrap_content"
    android:layout_height="wrap_content"
    android:text="@string/thirty"
    />
<CheckBox
    android:id="@+id/forty"
    android:layout_width="wrap_content"
    android:layout_height="wrap_content"
    android:text="@string/forty"
    />
```

```
    <CheckBox
        android:id="@+id/fifty"
        android:layout_width="wrap_content"
        android:layout_height="wrap_content"
        android:text="@string/fifty"
        />
</LinearLayout>
```

步骤03 编写Activity主文件：先声明10个全局变量，用于接收10个控件对象；在onCreate方法内，根据控件ID获得这10个对象并赋予相应的变量；编写监听器。打开"src\com.example.olympicgames\MainActivity.java"文件，修改并添加如下代码。

```java
package com.example.olympicgames;
import android.app.Activity;
import android.os.Bundle;
import android.widget.CheckBox;
import android.widget.CompoundButton;
import android.widget.CompoundButton.OnCheckedChangeListener;
import android.widget.RadioButton;
import android.widget.RadioGroup;
import android.widget.TextView;
import android.widget.Toast;

public class MainActivity extends Activity {
    //定义各控件的变量
    private TextView who = null;
    private TextView how = null;
    private RadioGroup who_group = null;
    private RadioButton china = null;
    private RadioButton america = null;
    private RadioButton others = null;
    private CheckBox less = null;
    private CheckBox thirty = null;
    private CheckBox forty = null;
    private CheckBox fifty = null;
    @Override
    public void onCreate(Bundle savedInstanceState) {
```

```java
        super.onCreate(savedInstanceState);
        setContentView(R.layout.activity_main);

        //获得对应的控件
        who = (TextView)findViewById(R.id.who);
        how = (TextView)findViewById(R.id.how);
        who_group = (RadioGroup)findViewById(R.id.who_group);
        china = (RadioButton)findViewById(R.id.china);
        america = (RadioButton)findViewById(R.id.america);
        others = (RadioButton)findViewById(R.id.others);
        less = (CheckBox)findViewById(R.id.less);
        thirty = (CheckBox)findViewById(R.id.thirty);
        forty = (CheckBox)findViewById(R.id.forty);
        fifty = (CheckBox)findViewById(R.id.fifty);

        //设置who_group的监听器,其实是一句代码,其参数是一个带有重构函数的对象
        who_group.setOnCheckedChangeListener(new RadioGroup.OnCheckedChangeListener() {
            public void onCheckedChanged(RadioGroup group, int checkedId) {
                // TODO Auto-generated method stub
                if(checkedId == china.getId()){
                    Toast.makeText(MainActivity.this,"中国", Toast.LENGTH_SHORT).show();
                }
                else if(checkedId == america.getId()){
                    Toast.makeText(MainActivity.this, "美国", Toast.LENGTH_SHORT).show();
                }
                else if(checkedId == others.getId()){
                    Toast.makeText(MainActivity.this, "其他国家", Toast.LENGTH_SHORT).show();
                }
            }
        });

        //下面为4个复选框的多选按钮分别建立监听器。首先是less的监听器
        less.setOnCheckedChangeListener(new OnCheckedChangeListener() {
            public void onCheckedChanged(CompoundButton buttonView, boolean isChecked) {
                // TODO Auto-generated method stub
                if(isChecked)
                {
```

```java
            Toast.makeText(MainActivity.this, "30个以下", Toast.LENGTH_SHORT).show();
        }
        else{
            Toast.makeText(MainActivity.this, "不是30个以下", Toast.LENGTH_SHORT).show();
        }
    }
});


//为thirty建立监听器
thirty.setOnCheckedChangeListener(new CompoundButton.OnCheckedChangeListener() {
    public void onCheckedChanged(CompoundButton buttonView, boolean isChecked) {
        // TODO Auto-generated method stub
        if(isChecked)
        {
            Toast.makeText(MainActivity.this, "30~39", Toast.LENGTH_SHORT).show();
        }
        else{
            Toast.makeText(MainActivity.this, "不是30~39", Toast.LENGTH_SHORT).show();
        }
    }
});

//为forty建立监听器
forty.setOnCheckedChangeListener(new OnCheckedChangeListener() {
    public void onCheckedChanged(CompoundButton buttonView, boolean isChecked) {
        // TODO Auto-generated method stub
        if(isChecked)
        {
            Toast.makeText(MainActivity.this, "40~49", Toast.LENGTH_SHORT).show();
        }
        else{
            Toast.makeText(MainActivity.this, "不是40~49", Toast.LENGTH_SHORT).show();
        }
    }
});
```

```
//为fifty建立监听器
fifty.setOnCheckedChangeListener(new OnCheckedChangeListener() {
    public void onCheckedChanged(CompoundButton buttonView, boolean isChecked) {
        // TODO Auto-generated method stub
        if(isChecked)
        {
            Toast.makeText(MainActivity.this, "50以上", Toast.LENGTH_SHORT).show();
        }
        else{
            Toast.makeText(MainActivity.this, "不是50以上", Toast.LENGTH_SHORT).show();
        }
    }
});
}
}
```

4.4 列表视图（ListView）

列表视图（ListView）是Android软件开发中非常重要的组件之一，它以列表的形式展示具体内容（如联系人等），并且能够根据数据的长度自适应显示，每个软件基本上都会用到列表视图。列表的显示需要3个元素：列表视图、适配器和数据。

（1）列表视图。

列表视图是指用于展示列表的视图。

（2）适配器。

适配器是指把数据映射到列表视图上的中介。适配器一般有3种：ArrayAdapter、SimpleAdapter和SimpleCursorAdapter，其中，以ArrayAdapter最为简单，只能展示一行字；SimpleAdapter有最好的扩充性，可以自定义各种效果；SimpleCursorAdapter可以认为是SimpleAdapter与数据库的简单结合，可以很方便地把数据库的内容以列表的形式展示出来。

（3）数据。

数据是指具体的将被映射的字符串、图片或者基本组件等。

■4.4.1 简单的列表视图

在列表中可以直接通过new ArrayAdapter()操作语句来绘制列表，但是，比较复杂的列表就需要使用自定义布局来实现了。

【案例4-5】：使用ListView编写一个界面，当点击某条记录时，用Toast显示相关信息。显示效果如图4-5所示。

图 4-5　效果图

操作步骤如下。

打开"src\com.example.listview\MainActivity.java"文件，修改并添加如下代码。

package com.example.listview;
import android.os.Bundle;
import android.app.ListActivity;
import android.view.Menu;
import android.view.View;
import android.widget.AdapterView;
import android.widget.AdapterView.OnItemClickListener;
import android.widget.ArrayAdapter;
import android.widget.ListView;
import android.widget.Toast;

public class MainActivity extends ListActivity {
　　private String[] mListStr = {"姓名：小王","性别：男","年龄：25","居住地：杭州","邮箱：××××@gmail.com","联系方式:1575718××××"};
　　ListView mListView = null;
　　@Override
　　protected void onCreate(Bundle savedInstanceState) {

```
            mListView = getListView();
            setListAdapter(new ArrayAdapter<String>(this,
                android.R.layout.simple_list_item_1, mListStr));
            mListView.setOnItemClickListener(new OnItemClickListener() {
              @Override
              public void onItemClick(AdapterView<?> adapterView, View view, int position,
                  long id) {
                Toast.makeText(MainActivity.this,"您选择了" + mListStr[position], Toast.LENGTH_SHORT).show();
              }
            });
            super.onCreate(savedInstanceState);
        }
    }
```

4.4.2 带标题的ListView列表

使用SimpleAdapter时需要注意，要用Map<String,Object> item来保存列表中每一项要显示的标题与文本，new SimpleAdapter操作的时候将map中的数据写入，程序就可以自动绘制列表了。

【案例4-6】：编写一个带标题的ListView列表，显示效果如图4-6所示。

图 4-6 带标题的 ListView 效果图

操作步骤如下。

打开"src\com.example.listviewother\MainActivity.java"文件,修改并添加如下代码。

```java
package com.example.listviewother;
import java.util.ArrayList;
import java.util.HashMap;
import java.util.Map;
import android.os.Bundle;
import android.app.Activity;
import android.view.Menu;
import android.os.Bundle;
import android.app.ListActivity;
import android.view.View;
import android.widget.AdapterView;
import android.widget.AdapterView.OnItemClickListener;
import android.widget.ArrayAdapter;
import android.widget.ListView;
import android.widget.SimpleAdapter;
import android.widget.Toast;

public class MainActivity extends ListActivity {
    private String[] mListTitle = { "姓名","性别","年龄","居住地","邮箱","手机号码"};
    private String[] mListStr = { "小胡","男","19","杭州",
        "xiaoXX@gmail.com" ,"1575718XXXX"};
    ListView mListView = null;
    ArrayList<Map<String,Object>> mData= new ArrayList<Map<String,Object>>();;

    @Override
    protected void onCreate(Bundle savedInstanceState) {
        mListView = getListView();
        int lengh = mListTitle.length;
        for(int i =0; i < lengh; i++) {
            Map<String,Object> item = new HashMap<String,Object>();
            item.put("title", mListTitle[i]);
            item.put("text", mListStr[i]);
            mData.add(item);
        }
        SimpleAdapter adapter = new SimpleAdapter(this,mData,android.R.layout.simple_list_item_2,
```

```
            new String[]{"title","text"},new int[]{android.R.id.text1,android.R.id.text2});
        setListAdapter(adapter);
        mListView.setOnItemClickListener(new OnItemClickListener() {
            @Override
            public void onItemClick(AdapterView<?> adapterView, View view, int position,
                long id) {
                Toast.makeText(MainActivity.this,"您选择了: " + mListTitle[position] + "内容: "+mListStr
[position], Toast.LENGTH_LONG).show();
            }
        });
        super.onCreate(savedInstanceState);
    }
}
```

4.4.3 带图片的ListView列表

由于SimpleAdapter类中的构造函数无法实现带图片的ListView列表的界面布局，所以必须自己写布局。使用Map<String,Object> item可保存列表中每一项需要显示的内容，如图片、标题、内容等。

【案例4-7】：编写一个带图片的ListView列表，显示效果如图4-7所示。

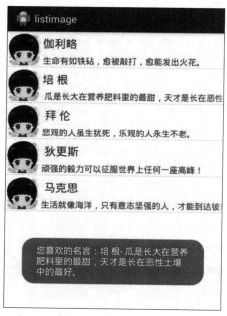

图 4-7 带图片的 ListView 列表的效果图

操作步骤如下。

步骤 01 编写"activity_main.xml"布局文件，添加1个ImageView控件和2个TextView控件。打开"res\layout\activity_main.xml"文件，修改并添加如下代码。

```xml
<RelativeLayout xmlns:android="http://schemas.android.com/apk/res/android"
    android:layout_width="fill_parent" android:layout_height="?android:attr/listPreferredItemHeight">
    <ImageView android:id="@+id/tupian"
        android:layout_width="wrap_content"
        android:layout_height="fill_parent"
        android:layout_alignParentTop="true"
        android:layout_alignParentBottom="true"
        android:adjustViewBounds="true"
        android:padding="2dip" />
    <TextView android:id="@+id/biaoti"
        android:layout_width="wrap_content"
        android:layout_height="wrap_content"
        android:layout_toRightOf="@+id/tupian"
        android:layout_alignParentRight="true"
        android:layout_alignParentTop="true"
        android:layout_above="@+id/wenzi"
        android:layout_alignWithParentIfMissing="true"
        android:gravity="center_vertical"
        android:textSize="20dip" />
    <TextView android:id="@+id/wenzi"
        android:layout_width="fill_parent"
        android:layout_height="wrap_content"
        android:layout_toRightOf="@+id/tupian"
        android:layout_alignParentBottom="true"
        android:layout_alignParentRight="true"
        android:singleLine="true"
        android:ellipsize="marquee"
        android:textSize="15dip" />
</RelativeLayout>
```

步骤 02 打开 "src\com.example.listimage\MainActivity.java" 文件，修改并添加如下代码。

```java
package com.example.listimage;
import java.util.ArrayList;
import java.util.HashMap;
import java.util.Map;
import android.os.Bundle;
import android.app.Activity;
import android.view.Menu;
import android.os.Bundle;
import android.app.ListActivity;
import android.view.View;
import android.widget.AdapterView;
import android.widget.AdapterView.OnItemClickListener;
import android.widget.ArrayAdapter;
import android.widget.ListView;
import android.widget.SimpleAdapter;
import android.widget.Toast;
public class MainActivity extends ListActivity {
    private String[] mListTitle = { "伽利略","培 根"," 拜 伦"," 狄更斯","马克思 "};
    private String[] mListStr = {"生命有如铁砧，愈被敲打，愈能发出火花。","瓜是长大在营养肥料里的最甜，天才是长在恶性土壤中的最好。","悲观的人虽生犹死，乐观的人永生不老。","顽强的毅力可以征服世界上任何一座高峰！","生活就像海洋，只有意志坚强的人，才能到达彼岸。"};
    ListView mListView = null;
    ArrayList<Map<String,Object>> mData= new ArrayList<Map<String,Object>>();
    @Override
    protected void onCreate(Bundle savedInstanceState) {
        mListView = getListView();
        int lengh = mListTitle.length;
        for(int i =0; i < lengh; i++) {
            Map<String,Object> item = new HashMap<String,Object>();
            item.put("image", R.drawable.one);
            item.put("title", mListTitle[i]);
            item.put("text", mListStr[i]);
            mData.add(item);
        }
        SimpleAdapter adapter = new SimpleAdapter(this,mData,R.layout.activity_main,
            new String[]{"image","title","text"},new int[]{R.id.tupian,R.id.biaoti,R.id.wenzi});
```

```
        setListAdapter(adapter);
        mListView.setOnItemClickListener(new OnItemClickListener() {
            @Override
            public void onItemClick(AdapterView<?> adapterView, View view, int position, long id) {
                Toast.makeText(MainActivity.this,"您喜欢的名言： " + mListTitle[position] + "-"+mListStr[position], Toast.LENGTH_SHORT).show();
            }
        });
        super.onCreate(savedInstanceState);
    }
}
```

4.5 网格视图（GridView）

网格视图是按照行、列的方式来显示内容的，一般用于显示图片等内容。例如，实现九宫格图，网格视图是首选，也是最简单的方式。在使用GridView时，经常用Adapter为GridView提供数据来源。网格视图应用中常用的对象及函数有如下几种。

（1）Context。

Context提供了关于应用环境全局信息的接口，它是一个抽象类，它的执行由Android系统提供，它允许获取以应用为特征的资源和类型，同时启动应用级的操作，如启动Activity、broadcasting和接收intents等。

（2）public void setAdapter (ListAdapter adapter)。

设置网格视图的数据。参数adapter为提供数据的适配器。

（3）public View getView(int position, View convertView, ViewGroup parent)。

该方法用来获得指定位置要显示的View，各参数的含义如下所述。

- **position**：该视图在适配器数据中的位置。
- **convertView**：旧视图。
- **parent**：此视图最终会被附加到的父级视图。

（4）ImageView。

ImageView可以显示任意图像，如图标等。ImageView类可以加载各种来源的图片（如资源或图片库等），需要计算出图像的尺寸，它可以在其他布局中使用并提供各种显示选项，如缩放和着色（渲染）等。

（5）public void setAdjustViewBounds (boolean adjustViewBounds)。

当需要在ImageView中调整边框并保持可绘制对象的比例时，将该值设为true。

（6）public void setScaleType (ImageView.ScaleType scaleType)。

控制图像应该如何缩放和移动，以使图像与ImageView一致。参数scaleType表示缩放的方式。

【案例4-8】：使用网格视图编写一个界面，效果如图4-8所示。

图4-8 网格视图效果图

操作步骤如下。

步骤01 编写"activity_main.xml"布局文件，添加一个GridView控件。打开"res\layout\activity_main.xml"文件，修改并添加如下代码。

```
<LinearLayout xmlns:android="http://schemas.android.com/apk/res/android"
  xmlns:tools="http://schemas.android.com/tools"
  android:layout_width="fill_parent"
  android:layout_height="fill_parent"
  android:orientation="vertical" >

  <GridView
    android:id="@+id/GridViewone"
    android:layout_width="wrap_content"
    android:layout_height="wrap_content" >
  </GridView>

</LinearLayout>
```

步骤02 打开"src\com.example.gridview\MainActivity.java"文件，修改并添加如下代码。

```
package com.example.gridview;
import android.os.Bundle;
import android.app.Activity;
import android.content.Context;
```

```java
import android.view.Menu;
import android.view.View;
import android.view.ViewGroup;
import android.widget.BaseAdapter;
import android.widget.GridView;
import android.widget.ImageView;

public class MainActivity extends Activity {
    private GridView gv;

    @Override
    public void onCreate(Bundle savedInstanceState) {
        super.onCreate(savedInstanceState);
        setContentView(R.layout.activity_main);
        gv=(GridView)findViewById(R.id.GridViewone);
        //设置GridView的列数
        gv.setNumColumns(3);
        //为GridView设置适配器
        gv.setAdapter(new MyAdapter(this));
    }
    //自定义适配器
    class MyAdapter extends BaseAdapter{
        //图片ID数组
        private Integer[] imgs = {
                R.drawable.one,
                R.drawable.two,
                R.drawable.three,
                R.drawable.four,
                R.drawable.five,
                R.drawable.six,
                R.drawable.seven,
                R.drawable.eight,
                R.drawable.nine,
        };
        //上下文对象
         Context context;
        //构造方法
```

```java
MyAdapter(Context context){
    this.context = context;
}
//获得数量
public int getCount() {
    // TODO Auto-generated method stub
    return imgs.length;
}
//获得当前选项
public Object getItem(int position) {
    // TODO Auto-generated method stub
    return position;
}
//获得当前选项ID
public long getItemId(int position) {
    // TODO Auto-generated method stub
    return position;
}
//创建View方法
public View getView(int position, View convertView, ViewGroup parent) {
    // TODO Auto-generated method stub
    ImageView imageView;
    if (convertView == null) {
        //实例化ImageView对象
        imageView = new ImageView(context);
        //设置ImageView对象布局
        imageView.setLayoutParams(new GridView.LayoutParams(125, 125));
        //设置边界对齐
        imageView.setAdjustViewBounds(false);
        //设置刻度类型
        imageView.setScaleType(ImageView.ScaleType.CENTER_CROP);
        //设置间距
        imageView.setPadding(8, 8, 8, 8);
    } else {
        imageView = (ImageView) convertView;
    }
    //为ImageView设置图片资源
```

```
                imageView.setImageResource(imgs[position]);
                return imageView;
            }
        }


@Override
public boolean onCreateOptionsMenu(Menu menu) {
    // Inflate the menu; this adds items to the action bar if it is present.
    getMenuInflater().inflate(R.menu.main, menu);
    return true;
}

}
```

课后作业

一、填空题

1. 在ImageButton中设置要显示的图标，可使用_____和_____方法。
2. 在ImageView控件中用于表示图片显示方式的属性是"_____"。
3. 每个软件基本上都会使用ListView。使用ListView需要3个元素：_____、_____、_____。

二、选择题

1. 若要实现九宫格图，在Android Studio中需要使用（   ）视图。

    A. ListView

    B. Spinner

    C. GridView

    D. DatePicker

2. 不属于列表显示元素的是（   ）。

    A. 列表视图

    B. 适配器

    C. 数据

    D. 网格视图

3. 在Android中基础图片的显示控件是（　　）。

   A. ImageView

   B. ImageButton

   C. CheckBox

   D. ListView

4. android:checked属性是（　　）控件的特有属性。

   A. CheckBox

   B. ImageButton

   C. RadioButton

   D. ListView

三、操作题

列表视图是Android软件开发中非常重要的组件之一，它以列表的形式展示具体内容。利用Android Studio开发环境创建一个新的Android应用程序，实现一个简单的纯文本列表控件和一个相对复杂的列表控件。（提示：参照书中4.4节内容进行练习。）

第 5 章

多线程与事件处理机制

内容概要

在设计应用程序时,如何给用户以舒适的体验一直是开发人员追求的目标。多线程机制的使用可以有效加强应用程序处理状态与用户之间的互动,提升用户体验效果。另外,为了高效应对用户指令,Android提供了强大的事件处理机制。本章将介绍有关多线程与事件处理机制的相关知识。

数字资源

【本章案例文件】:"案例文件\第5章"目录下

5.1 Android的多线程

线程是进程中的一个实体，是被系统独立调度和分派的基本单位。线程本身不拥有系统资源，或只拥有少量运行中必不可少的资源，但它可与同属一个进程的其他线程共享进程所拥有的全部资源。一个线程可以创建和撤销另一个线程，同一进程中的多个线程之间可以并发执行。线程之间相互制约，在运行中呈现间断性。线程有就绪、阻塞和运行3种基本状态。每一个程序都至少有一个线程，若程序只有一个线程，那就是程序本身。线程是程序中一个单一的顺序控制流程。在单个程序中同时运行多个线程完成不同的工作，称为多线程。

为什么要使用多线程呢？因为通过多线程可以提高资源的使用效率和系统的执行效率。提升效率主要体现在以下几点。

- 使用线程可以合理分配"时间片"，把占用时间长的任务放置到后台处理。
- 实现更人性化的用户界面设计。例如，当用户点击按钮触发了某些事件的处理，会弹出一个进度条来显示处理的进度。
- 在一些需要等待的任务实现上，如用户输入、文件读写和网络收发数据等，利用多线程可以释放一些资源。

■5.1.1 多线程机制的优缺点

Android的多线程是基于Linux本身的多线程机制，而且多线程之间的同步又是通过Java本身的线程同步实现的。在Android中使用多线程的主要优势体现在以下3个方面。

1. 避免应用程序无响应，提升用户体验

在Android中，如果在某段时间内应用程序响应不够灵敏，系统会向用户弹出一个对话框，这个对话框称为应用程序无响应（Application Not Responding，ANR）对话框。用户可以选择"等待"而让程序继续运行，也可以选择"强制关闭"，停止运行程序。默认情况下，在Android中Activity的最长执行时间是5 s，BroadcastReceiver的最长执行时间则是10 s。一个运行流畅的应用程序不应该出现程序无响应的情况，因为这会带来令人不满意的体验，使用多线程则可以避免程序无响应的情况。例如，在访问网络服务端时返回过慢、数据过多而导致滑动屏幕不流畅，或者I/O读取大资源时，可通过开启一个新线程来处理这些耗时的操作。开发程序的一个事件处理原则是把所有可能耗时的操作都放到其他线程去处理。参考代码如下。

```
new Thread() {
    public void run() {
    //耗时操作代码
    }
}.start();
```

Android中的Main线程在时间处理上，若在5 s内无法得到响应，就会弹出程序无响应对话框。在Main线程中执行的方法有两种。

- 使用Activity的生命周期方法，如onCreate、onStart、onResume等方法。
- 使用事件处理方法，如onClick、onItemClick等方法。

一般来说，Activity的onCreate、onStart、onResume方法的执行时间决定了应用首页打开的时间，所以应尽量把不必要的操作放到其他线程去处理；如果仍然很耗时，可以使用动态的应用图标告知用户应用正在运行。

2. 实现异步处理

当用户与应用程序交互时，事件处理方法的执行效率决定了应用程序的响应性能。事件处理方法分为同步和异步两种。

同步，需要等待返回结果。例如，用户点击了"登录"按钮，需要等待服务端返回结果，此时需要有一个进度条来提示用户程序的运行情况。

异步，不需要等待返回结果。异步的概念和同步相对。当一个异步进程调用发出后，调用者不能立刻得到结果，调用的部件在完成后通过状态、通知和回调来通知调用者。例如，微博中的收藏功能，点击"收藏"按钮后，会通知"收藏成功"而无须用户等待，这里就需要使用异步方式实现。

无论同步还是异步，事件处理都有可能比较耗时，此时就需要将此事件放到其他线程中处理，处理完成后再通知刷新界面。需要注意的是，不是所有的界面刷新行为都需要放到Main线程中处理。例如，TextView的setText方法需要在Main线程中处理，否则会抛出CalledFromWrong-ThreadException异常；而ProgressBar的setProgress方法则不需要在Main线程中处理。

3. 实现多任务

多任务是指一个操作系统可以同时执行多个程序的能力。基本上，操作系统使用一个硬件时钟为同时运行的每个进程分配"时间片"。如果时间片足够小，并且机器负荷不重，那么在用户看来，所有的程序似乎在"同时"运行。应用程序使用多线程在后台执行长作业，此时用户仍然可以使用计算机进行其他工作。例如，向打印机发出打印的命令，假如此时计算机停止响应了，就必须停止手上的工作来等待低速的打印机工作吗？一般情况是，在打印机工作的同时还可以使用听音乐、画图、看电影等各种应用程序。这是因为每一个程序被分成了独立的不同的任务，使用多线程，即使某一部分任务失败了，也不会对其他任务造成影响，不会导致整个程序崩溃。

总之，使用多线程可以获得更高的CPU利用率、更好的系统可靠性，并改善多处理器计算机的性能等。在许多应用中，可以同步调用资源。但即使多线程具有以上诸多的优势，也不可过多使用多线程。因为过多使用多线程会出现数据同步的问题，这需要特别处理，在使用多线程的时候必须尽量保证每个线程的独立性不会被其他线程干预。另外，如果一个程序有多个线程，则其他程序的线程必然只能占用更少的CPU时间，因为还需要大量的CPU时间做线程调度，用大量操作系统的内存空间维护每个线程的上下文信息，这样反而会降低系统的运行效率。

5.1.2 多线程的实现

Android中的线程是基于Java定义的线程。一个应用程序中可能会包含多个线程（thread），每个线程中都有一个run方法，run方法内部的程序执行完毕后，所在的线程就自动结束。每个线程都有一个消息队列，用于在不同线程之间传递消息。

1. 线程定义

Android中定义线程的方法和Java是一样的，有两种方式，一种是Thread，另一种是Runnable。Thread是一个类，根据Java的继承要求，一个类只能有一个父类，所以继承了Thread的子类就不能再继承其他类，因而限制了这种方法的使用。Runnable是一个接口（interface），同样可以启动一个线程，不同的是它可以被多继承。使用Thread类的方式实现线程的实例代码如下。

```java
package org.thread.demo;
class MyThread extends Thread{
 private String name;
 public MyThread(String name) {
    super();
    this.name = name;
 }
 public void run(){
    for(int i=0;i<10;i++){
    System.out.println("线程开始： "+this.name+",i="+i);
    }
 }
}
package org.thread.demo;
public class ThreadDemo01 {
 public static void main(String[] args) {
    MyThread mt1=new MyThread("线程a");
    MyThread mt2=new MyThread("线程b");
    mt1.run();
    mt2.run();
 }
}
```

程序的运行很有规律，先执行第1个对象，然后执行第2个对象，不会交互运行。在JDK的文档中可以发现，一旦调用start方法，则会通过JVM找到run方法。下面用start方法启动线程，代码如下。

```java
package org.thread.demo;
public class ThreadDemo01 {
public static void main(String[] args) {
   MyThread mt1=new MyThread("线程a");
   MyThread mt2=new MyThread("线程b");
   mt1.start();
   mt2.start();
}
}
```

这样程序就可以正常完成交互式运行。为什么要使用start方法启动多线程呢？这是因为在JDK的安装路径下，"src.zip"文件是全部的Java源程序，通过此代码能找到Thread中的start方法的定义，可以发现此方法中使用了private native void start0();，其中的native关键字表示可以调用操作系统的底层函数，这样的技术称为JNI技术（Java Native Interface）。

在实际开发中一个多线程的操作很少使用Thread类，而大多是通过Runnable接口完成的。

```java
public interface Runnable{
public void run();
}
```

例如，使用Runnable接口实现线程的实例代码如下。

```java
package org.runnable.demo;
class MyThread implements Runnable{
private String name;
public MyThread(String name) {
   this.name = name;
}
public void run(){
   for(int i=0;i<100;i++){
   System.out.println("线程开始： "+this.name+",i="+i);
   }
}
}
```

但是，在使用Runnable定义的子类中没有start方法，只有Thread类中才有。Thread类中有一个构造方法public Thread(Runnable targer)，此构造方法中接收Runnable的子类实例，也就是说可以通过Thread类来启动Runnable接口实现多线程。Thread类中的start方法可以协调系统的资源，示例代码如下。

```
package org.runnable.demo;
import org.runnable.demo.MyThread;
public class ThreadDemo01 {
    public static void main(String[] args) {
        MyThread mt1=new MyThread("线程a");
        MyThread mt2=new MyThread("线程b");
        new Thread(mt1).start();
        new Thread(mt2).start();
    }
}
```

在实际程序开发中多线程的实现多以Runnable接口为主,因为实现Runnable接口比继承Thread类有一些好处:可避免类继承的局限,一个类可以继承多个接口;适合资源的共享等。

2. Handler、Message和Looper

Handler主要接收子线程发送的数据,并用此数据配合主线程更新UI。一个线程中只能有一个Handler对象,可以通过该对象向所在线程发送消息。Handler主要有两种用途:一是实现定时任务,类似于Windows程序中的定时器功能,可以通过Handler对象向所在线程发送一个延时消息,当线程运行到达指定的时间后,则通过Handler的消息处理函数完成指定的任务;二是在线程间传递数据。

(1) 完成定时任务。

在一个Activity内部实现定时器功能,需要通过Handler对象发送延迟消息的方法实现。Handler有两种发送消息的方式:一种是postXXX方法,用于把一个Runnable对象发送到消息队列,当消息被处理时,能够执行Runnable对象;另一种是sendXXX方法,用于发送一个Message类型的消息到消息队列,当消息被处理时,系统会调用Handler对象定义的handleMessage()方法处理该消息。

sendXXX方法包含以下方法。

- **sendEmptyMessage(int)**:发送空消息。
- **sendMessage(Message)**:发送Message指定的消息。
- **sendMessageAtTime(Message,long)**:在指定的时间点发送Message指定的消息。
- **sendMessageDelayed(Message,long)**:在指定的时间后发送Message指定的消息。

(2) 在线程之间传递数据。

如果一个进程获得了另一个进程的Handler,那么这个进程就可以通过handler.sendMessage(Message)方法向那个进程发送数据。基于此机制,在处理多线程的时候可以新建一个Thread,这个Thread拥有UI线程中的一个Handler。当Thread处理完一些耗时的操作后,通过传递过来的Handler向UI线程发送数据,由UI线程去更新界面。

线程在默认的情况下,只要run()函数执行完毕,线程就结束,但有时新建的线程需要接收消息并处理。因此，在新线程中，除了需要添加一个Handler对象外，还需要从线程的消息队列中取出消息，并负责分发消息，这就需要用Looper实现了。事实上，Activity内部只有一个Looper，只是Activity是一个特殊的线程，操作系统已经将其封装了而已。

在Android中，Handler和Message、Thread有密切的关系。Handler主要是负责Message的分发和处理。Message是由一个消息队列进行管理，消息队列又是由一个Looper进行管理的，即在Android系统中，Looper负责管理线程的消息队列和消息循环。通过Loop.myLooper()可以得到当前线程的Looper对象，通过Loop.getMainLooper()可以获得当前进程的主线程的Looper对象。Android系统的消息队列和消息循环都是针对具体线程的。一个线程可以存在（也可以不存在）一个消息队列和一个消息循环（Looper），特定线程的消息只能分发给本线程，不能进行跨线程、跨进程的通信。但是用户创建的工作线程默认是没有消息循环和消息队列的，如果想让该线程具有消息队列和消息循环，首先需要在线程中调用Looper.prepare()方法创建消息队列，然后调用Looper.loop()方法进入消息循环。

若线程中有一个Looper对象，其内部维护了一个消息队列。一个线程只能有一个Looper对象，Looper类的源代码如下。

```java
public class Looper {
    // 每个线程中的Looper对象其实是一个ThreadLocal，即线程本地存储(TLS)对象
    private static final ThreadLocal sThreadLocal = new ThreadLocal();
    // Looper内的消息队列
    final MessageQueue mQueue;
    // 当前线程
    Thread mThread;
    //其他属性
    // 每个Looper对象中有它的消息队列和它所属的线程
    private Looper() {
        mQueue = new MessageQueue();
        mRun = true;
        mThread = Thread.currentThread();
    }
    // 调用该方法会在调用线程的TLS中创建Looper对象
    public static final void prepare() {
        if (sThreadLocal.get() != null) {
            // 试图在有Looper的线程中再次创建Looper将抛出异常
            throw new RuntimeException("Only one Looper may be created per thread");
        }
        sThreadLocal.set(new Looper());
    }
}
```

调用下面的loop方法后，Looper线程就开始真正工作了，它不断从自己的消息队列中取出队头的消息（也称任务）执行。其源代码如下。

```java
public static final void loop() {
    Looper me = myLooper();  //得到当前线程的Looper
    MessageQueue queue = me.mQueue;  //得到当前Looper的MessageQueue
    Binder.clearCallingIdentity();
    final long ident = Binder.clearCallingIdentity();
    // 开始循环
    while (true) {
        Message msg = queue.next(); // 取出message
        if (msg != null) {
            if (msg.target == null) {
                // message没有target为结束信号，退出循环
                return;
            }
            // 日志
            if (me.mLogging!= null)
                me.mLogging.println(
                    ">>>>> Dispatching to " + msg.target + " " + msg.callback + ": " + msg.what
                );
            msg.target.dispatchMessage(msg);
            //将处理工作交给message的target，即handler
            if (me.mLogging!= null)
                me.mLogging.println(
                    "<<<<< Finished to    " + msg.target + " " + msg.callback);
            final long newIdent = Binder.clearCallingIdentity();
            if (ident != newIdent) {
                Log.wtf("Looper", "Thread identity changed from 0x"
                    + Long.toHexString(ident) + " to 0x"
                    + Long.toHexString(newIdent) + " while dispatching to "
                    + msg.target.getClass().getName() + " "
                    + msg.callback + " what=" + msg.what);
            }
            msg.recycle();  // 回收message资源
        }
    }
}
```

除了prepare和loop方法，Looper类还提供了3种方法：myLooper方法用于获得当前线程的Looper对象；getThread方法用于获得Looper对象所属的线程；quit方法用于结束循环。

Handler扮演了向消息队列中添加消息和处理消息的角色（只处理由自己发出的消息），即通知消息队列，它要执行一个任务，如sendMessage等，并在循环执行到自己的时候执行该任务，即handleMessage，整个过程是异步的。Handler创建时会关联一个Looper，默认的构造方法是关联当前线程的Looper，不过这也是可以设置的。LooperThread类加入Handler的实现代码如下。

```java
public class LooperThread extends Thread {
    private Handler handler1;
    private Handler handler2;
    @Override
    public void run() {
        // 将当前线程初始化为Looper线程
        Looper.prepare();
        // 实例化两个handler
        handler1 = new Handler();
        handler2 = new Handler();
        // 开始循环处理消息队列
        Looper.loop();
    }
}
```

一个线程可以有多个Handler，但是只能有一个Looper。Handler可以向任意线程发送消息：首先创建消息，然后根据Looper找到相关联的消息队列，将这些消息添加到关联的消息队列中；Handler是在它关联的Looper线程中处理消息的，Looper先取出消息队列的头消息，对应的Handler就会执行handlerMessage；最后返回Looper继续执行。这就解决了Android不能在其他非主线程中更新UI的问题。

Android的主线程也是一个Looper线程，在其中创建的Handler默认将关联主线程的消息队列。因此，利用Handler的一个解决方案就是在Activity中创建Handler，并将其引用传递给工作线程，工作线程执行完任务后会使用Handler发送消息，通知Activity更新UI。

3. 线程间的消息传递

一般情况下，特定线程的消息只能分发给本线程，不能进行跨线程的通信，但也不是绝对的。因为在Android中可以通过获得线程的Looper对象来实现不同线程间消息的传递，代码如下。

```java
package com.mytest.handlertest;
import android.app.Activity;
import android.graphics.Color;
```

```java
import android.os.Bundle;
import android.os.Handler;
import android.os.Looper;
import android.os.Message;
import android.util.Log;
import android.view.View;
import android.view.View.OnClickListener;
import android.view.ViewGroup.LayoutParams;
import android.widget.Button;
import android.widget.LinearLayout;
import android.widget.TextView;
public class HandlerTest extends Activity implements OnClickListener{
    private String TAG = "HandlerTest";
    private boolean bpostRunnable = false;
    private NoLooperThread noLooperThread = null;
    private OwnLooperThread ownLooperThread = null;
    private ReceiveMessageThread receiveMessageThread =null;
    private Handler  mOtherThreadHandler=null;
    private EventHandler mHandler = null;
    private Button btn1 = null;
    private Button btn2 = null;
    private Button btn3 = null;
    private Button btn4 = null;
    private Button btn5 = null;
    private Button btn6 = null;
    private TextView tv = null;
    /** Called when the activity is first created. */
    @Override
    public void onCreate(Bundle savedInstanceState) {
        super.onCreate(savedInstanceState);
        LinearLayout layout = new LinearLayout(this);
        LinearLayout.LayoutParams params = new LinearLayout.LayoutParams(250, 50);
        layout.setOrientation(LinearLayout.VERTICAL);
        btn1 = new Button(this);
        btn1.setId(101);
        btn1.setText("message from main thread self");
        btn1.setOnClickListener(this);
```

```
layout.addView(btn1, params);
btn2 = new Button(this);
btn2.setId(102);
btn2.setText("message from other thread to main thread");
btn2.setOnClickListener(this);
layout.addView(btn2,params);
btn3 = new Button(this);
btn3.setId(103);
btn3.setText("message to other thread from itself");
btn3.setOnClickListener(this);
layout.addView(btn3, params);
btn4 = new Button(this);
btn4.setId(104);
btn4.setText("message with Runnable as callback from other thread to main thread");
btn4.setOnClickListener(this);
layout.addView(btn4, params);
btn5 = new Button(this);
btn5.setId(105);
btn5.setText("main thread's message to other thread");
btn5.setOnClickListener(this);
layout.addView(btn5, params);
btn6 = new Button(this);
btn6.setId(106);
btn6.setText("exit");
btn6.setOnClickListener(this);
layout.addView(btn6, params);
tv = new TextView(this);
tv.setTextColor(Color.WHITE);
tv.setText("");
params = new LinearLayout.LayoutParams(LayoutParams.FILL_PARENT,LayoutParams.WRAP_CONTENT);
params.topMargin=10;
layout.addView(tv, params);
setContentView(layout);
receiveMessageThread = new ReceiveMessageThread();
receiveMessageThread.start();
}
```

```java
class EventHandler extends Handler{
    public EventHandler(Looper looper){
        super(looper);
    }

    public EventHandler(){
        super();
    }
    @Override
    public void handleMessage(Message msg) {
        // TODO Auto-generated method stub
        super.handleMessage(msg);
        Log.e(TAG, "CurrentThread id:----------+>" + Thread.currentThread().getId());
        switch(msg.what){
        case 1:
            tv.setText((String)msg.obj);
            break;
        case 2:
            tv.setText((String)msg.obj);
            noLooperThread.stop();
            break;
        case 3:
            //不能在非主线程的线程里面更新UI，所以这里通过Log打印信息
            Log.e(TAG,(String)msg.obj);
            ownLooperThread.stop();
            break;
        default:
            Log.e(TAG,(String)msg.obj);
            break;
        }
    }
}
//ReceiveMessageThread has his own message queue by execute Looper.prepare();
class ReceiveMessageThread extends Thread {
    @Override
    public void run(){
        Looper.prepare();
```

```java
        mOtherThreadHandler= new Handler(){
            @Override
            public void handleMessage(Message msg) {
                // TODO Auto-generated method stub
                super.handleMessage(msg);
                Log.e(TAG,"-------+>"+(String)msg.obj);
                Log.e(TAG, "CurrentThread id:----------+>" + Thread.currentThread().getId());
            }
        };
        Log.e(TAG, "ReceiveMessageThread id:--------+>" + this.getId());
        Looper.loop();
    }
}

class NoLooperThread extends Thread {
    private EventHandler mNoLooperThreadHandler;
    @Override
    public void run() {
        Looper myLooper = Looper.myLooper();
        Looper mainLooper= Looper.getMainLooper();
        String msgobj;
        if(null == myLooper){
//这里获得的是主线程的Looper，由于NoLooperThread没有自己的Looper，所以这里肯定会被执行
            mNoLooperThreadHandler = new EventHandler(mainLooper);
            msgobj = "NoLooperThread has no looper and handleMessage function executed in main thread!";
        } else{
            mNoLooperThreadHandler = new EventHandler(myLooper);
            msgobj = "This is from NoLooperThread self and handleMessage function executed in NoLooperThread!";
        }
        mNoLooperThreadHandler.removeMessages(0);
        if(bpostRunnable == false){
//send message to main thread
            Message msg = mNoLooperThreadHandler.obtainMessage(2, 1, 1, msgobj);
            mNoLooperThreadHandler.sendMessage(msg);
            Log.e(TAG, "NoLooperThread id:--------+>" + this.getId());
        }else{
```

```java
        //下面new Runnable接口的对象中的run()函数是在Main Thread中执行，不是在
        //NoLooperThread中执行
        //注意，Runnable是一个接口，它里面的run()函数被执行时不会再新建一个线程
        mNoLooperThreadHandler.post(new Runnable(){
            public void run() {
                // TODO Auto-generated method stub
                tv.setText("update UI through handler post runnalbe mechanism!");
                Log.e(TAG, "update UI id:--------+>" + Thread.currentThread().getId());
                noLooperThread.stop();
            }
        }
        );
    }
}
class OwnLooperThread extends Thread{
    private EventHandler mOwnLooperThreadHandler = null;
    @Override
    public void run() {
        Looper.prepare();
        Looper myLooper = Looper.myLooper();
        Looper mainLooper= Looper.getMainLooper();
        String msgobj;
        if(null == myLooper){
            mOwnLooperThreadHandler = new EventHandler(mainLooper);
            msgobj = "OwnLooperThread has no looper and handleMessage function executed in main thread!";
        }
        else{
            mOwnLooperThreadHandler = new EventHandler(myLooper);
            msgobj = "This is from OwnLooperThread self and handleMessage function executed in NoLooperThread!";
        }
        mOwnLooperThreadHandler.removeMessages(0);
        //给自己发送消息
        Message msg = mOwnLooperThreadHandler.obtainMessage(3,1,1,msgobj);
        mOwnLooperThreadHandler.sendMessage(msg);
        Looper.loop();
    }
```

```java
}
public void onClick(View v) {
    // TODO Auto-generated method stub
    switch(v.getId()){
    case 101:
        //主线程发送消息给自己
        Looper looper = Looper.myLooper();//get the Main looper related with the main thread
        //如果不给任何参数，会用当前线程对应的Looper（这里就是Main Looper）为Handler里面
        //的成员mLooper赋值
        mHandler = new EventHandler(looper);
        // 清除整个MessageQueue里的消息
        mHandler.removeMessages(0);
        String obj = "This main thread's message and received by itself!";
        Message msg = mHandler.obtainMessage(1,1,1,obj);
        // 将Message对象送入主线程的MessageQueue里面
        mHandler.sendMessage(msg);
        break;
    case 102:
        //other线程发送消息给主线程
        bpostRunnable = false;
        noLooperThread = new NoLooperThread();
        noLooperThread.start();
        break;
    case 103:
        //other线程获取它自己发送的消息
        tv.setText("please look at the error level log for other thread received message");
        ownLooperThread = new OwnLooperThread();
        ownLooperThread.start();
        break;
    case 104:
        //other线程通过Post Runnable方式发送消息给主线程
        bpostRunnable = true;
        noLooperThread = new NoLooperThread();
        noLooperThread.start();
        break;
    case 105:
        //主线程发送消息给other线程
```

```
            if(null!=mOtherThreadHandler){
            tv.setText("please look at the error level log for other thread received message from main thread");
            String msgObj = "message from mainThread";
            Message mainThreadMsg = mOtherThreadHandler.obtainMessage(1, 1, 1, msgObj);
            mOtherThreadHandler.sendMessage(mainThreadMsg);
            }
            break;
        case 106:
            finish();
            break;
        }
    }
}
```

5.2 事件处理机制

Android系统提供了两套强大的事件处理机制：一套是基于监听的事件处理机制，另一套是基于回调的事件处理机制。Android基于监听的事件处理，主要的做法是为Android界面组件绑定特定的事件监听器。Android对于回调的事件处理，主要的做法是重载Android组件特定的回调方法或者重载Activity。

■5.2.1 基于监听接口的事件处理

对于Android的应用程序来说，事件处理是必不可少的。用户与应用程序之间的交互便是通过事件处理来完成的。关于Android事件处理应该注意以下两点。

（1）事件源与事件监听器。当用户与应用程序交互时，一定是通过触发某些事件完成的，由事件通知程序应该执行哪些操作，在这个过程中主要涉及事件源与事件监听器两个对象。

- 事件源指的是事件所发生的控件，各个控件在不同情况下触发的事件不尽相同，而且产生的事件对象也可能不同。
- 监听器是用来处理事件的对象，它实现了特定的接口。用户需要根据事件不同重载不同的事件。

（2）将事件源与事件监听器联系到一起，就需要为事件源注册监听器。当事件发生时，系统才会自动通知事件监听器来处理相应的事件。

事件处理的过程一般分为3步。

（1）为事件源对象添加监听器，这样当某个事件被触发时，系统才会知道通知谁来处理该事件。

（2）当事件发生时，系统会将事件封装成相应类型的事件对象，并发送给注册到事件源的事件监听器。

（3）当监听器对象接收到事件对象之后，系统会调用监听器中相应的事件处理方法来处理事件并给出响应。

主要的监听器接口说明如下。

1. OnClickListener接口

（1）功能。

OnClickListener接口处理的是点击事件。在触控模式下，是指在某个View上按下并抬起的组合动作；而在键盘模式下，是某个View获得焦点后点击确定键或者按下轨迹球的事件。

（2）回调方法。

public void onClick(View v)

说明：参数v是事件发生的事件源。

2. OnLongClickListener接口

（1）功能。

OnLongClickListener接口与OnClickListener接口的原理基本相同，只是该接口为View长按事件的捕捉接口，即当长时间按下某个View时触发的事件。

（2）回调方法。

public boolean onLongClick(View v)

说明：参数v为事件源控件，在长时间按下此控件时才会触发该方法。

返回值：该方法的返回值为一个boolean类型的变量。当返回值为true时，表示已经完整地处理了这个事件，并不希望其他的回调方法再次进行处理；当返回值为false时，表示并没有完全处理完该事件，更希望其他方法继续对其进行处理。

3. OnFocusChangeListener接口

（1）功能。

OnFocusChangeListener接口用于处理控件焦点发生改变的事件。如果注册了该接口，当某个控件失去焦点或者获得焦点时都会触发该接口中的回调方法。

（2）回调方法。

public void onFocusChange(View v, Boolean hasFocus)

说明：参数v为触发该事件的事件源；参数hasFocus表示v的新状态，即v是否获得焦点。

4. OnKeyListener接口

（1）功能。

OnKeyListener是对手机键盘进行监听的接口，通过对某个View注册该监听，当View获得焦点并有键盘事件时，便会触发该接口中的回调方法。

（2）回调方法。

public boolean onKey(View v, int keyCode, KeyEvent event)

说明：参数v为事件的事件源控件；参数keyCode为手机键盘的键盘码；参数event是键盘事件封装类对象，其中包含了事件的详细信息，如发生的事件、事件的类型等。

5. OnTouchListener接口

（1）功能。

OnTouchListener接口是用于处理手机屏幕事件的监听接口，当在View的范围内触摸按下、抬起或滑动等动作时都会触发该事件。

（2）回调方法。

public boolean onTouch(View v, MotionEvent event)

说明：参数v为事件源对象；参数event为事件封装类对象，其中封装了触发事件的详细信息，包括事件的类型、触发时间等信息。

6. OnCreateContextMenuListener接口

（1）功能。

OnCreateContextMenuListener接口是用于处理上下文菜单显示事件的监听接口。该方法是定义和注册上下文菜单的另一种方式。

（2）回调方法。

public void onCreateContextMenu(ContextMenu menu, View v, ContextMenuInfo info)

说明：参数menu为事件的上下文菜单；参数v为事件源View，当该View获得焦点时才可能接收该方法的事件响应；参数info对象中封装了上下文菜单额外的信息，这些信息取决于事件源View。

■5.2.2 基于回调机制的事件处理

回调机制实质上就是将事件的处理绑定在控件上，由图形用户界面控件处理事件，回调机制需要自定义View来实现。Android平台中，每个View都有自己的处理事件的回调方法，开发人员可以通过重载View中的这些回调方法实现需要的响应事件。当某个事件没有被任何一个View处理时，便会调用Activity中相应的回调方法。主要的回调方法说明如下。

1. onKeyDown方法

（1）功能。

onKeyDown方法是接口KeyEvent.Callback中的抽象方法，该方法用于捕捉手机键盘被按下的事件。所有的View都实现了该接口并重载了该方法。

（2）方法声明。

public boolean onKeyDown (int keyCode, KeyEvent event)

- keyCode：该参数为被按下的键值，即键盘码。手机键盘中每个按钮都会有其单独的键盘码，应用程序都是通过键盘码知道用户按下的是哪个键。
- event：该参数为按键事件的对象，其中包含了触发事件的详细信息，如事件的状态、事件的类型、事件发生的时间等。当用户按下按键时，系统会自动将事件封装成KeyEvent对象供应用程序使用。
- 返回值：该方法的返回值为一个boolean类型的变量。当返回值为true时，表示已经完整地处理了这个事件，并不希望其他的回调方法再次进行处理；当返回值为false时，表示并没有完全处理完该事件，更希望其他回调方法继续对其进行处理，如Activity中的回调方法等。

2. onKeyUp方法

（1）功能。

onKeyUp方法同样是接口KeyEvent.Callback中的一个抽象方法，该方法用于捕捉手机键盘按键抬起的事件。所有的View都实现了该接口并重载了该方法。

（2）方法声明。

public boolean onKeyUp (int keyCode, KeyEvent event)

- keyCode：该参数为触发事件的按键码，需要注意的是，同一个按键在不同型号的手机中的按键码可能不同。
- event：该参数为事件封装类的对象，其含义与onKeyDown方法中的完全相同，在此不再赘述。
- 返回值：该方法的返回值表示的含义与onKeyDown方法相同，同样也是通知系统是否希望其他回调方法再次对该事件进行处理。

3. onTouchEvent方法

（1）功能。

onTouchEvent方法是在View类中定义的，并且所有的View子类都重载了该方法。应用程序通过该方法处理手机屏幕的触摸事件。

（2）方法声明。

public boolean onTouchEvent (MotionEvent event)

- event：该参数为手机屏幕触摸事件封装类的对象，其中封装了该事件的所有信息，如触摸的位置、触摸的类型、触摸的时间等。该对象会在用户触摸手机屏幕时被创建。
- 返回值：该方法的返回值机理与键盘响应事件相同，同样是当已经完整地处理了该事件

且不希望其他回调方法再次处理时返回true，否则返回false。

说明：该方法并不像前面两种方法一样只处理一种事件。一般情况下，以下3种情况下的事件都由onTouchEvent方法处理，只是3种情况中的动作值不同。

- 屏幕被按下：当屏幕被按下时，会自动调用该方法处理事件，此时MotionEvent.getAction()的值为MotionEvent.ACTION_DOWN。如果在应用程序中需要处理屏幕被按下的事件，只需重新调回该方法，然后在方法中进行动作的判断即可。
- 屏幕被抬起：当触控笔离开屏幕时触发的事件，该事件同样需要onTouchEvent方法来捕捉，然后在方法中进行动作判断。当MotionEvent.getAction()的值为MotionEvent.ACTION_UP时，表示是屏幕被抬起的事件。
- 在屏幕中拖动：该方法还负责处理触控笔在屏幕上滑动的事件，同样是调用MotionEvent.getAction()方法判断动作值是否为MotionEvent.ACTION_MOVE，然后再进行处理。

4. onTrackBallEvent方法

（1）功能。

onTrackBallEvent方法为手机中轨迹球的处理方法。同样所有的View都实现了该方法。

（2）方法声明。

```
public boolean onTrackballEvent (MotionEvent event)
```

- event：该参数为手机轨迹球事件封装类的对象，其中封装了触发事件的详细信息，包括事件的类型、触发时间等。一般情况下，该对象会在用户操控轨迹球时被创建。
- 返回值：该方法的返回值与前面介绍的各个回调方法的返回值机制完全相同。

5.2.3 回调方法应用案例

本节通过一个简单案例介绍onTouchEvent方法，其他回调方法的使用方式与onTouchEvent方法类似，读者可以自行练习。

【案例】：在用户点击的位置绘制一个矩形，然后监测用户触控笔的状态。当用户在屏幕上移动触控笔时，矩形随之移动；而当用户的触控笔离开手机屏幕时，则停止绘制矩形。

开发步骤如下。

创建一个名为"DrawRactangle"的Android项目。打开"DrawRactangle.java"文件，输入如下代码。

package wyf.ytl;	//声明所在包
import android.app.Activity;	//引入Activity类
import android.content.Context;	//引入Context类
import android.graphics.Canvas;	//引入Canvas类
import android.graphics.Color;	//引入Color类
import android.graphics.Paint;	//引入Paint类

```java
import android.os.Bundle;              //引入Bundle类
import android.view.MotionEvent;       //引入MotionEvent类
import android.view.View;              //引入View类
public class DrawRactangle extends Activity {
    MyView myView;                     //自定义View的引用
    public void onCreate(Bundle savedInstanceState)
    {
    //重载onCreate方法，该方法会在此Activity创建时被系统调用，在方法中先初始化自定义
    //的View，然后将当前的用户界面设置成该View
    super.onCreate(savedInstanceState);
    myView = new MyView(this);         //初始化自定义的View
    setContentView(myView);            //设置当前显示的用户界面
    }
    @Override
    public boolean onTouchEvent(MotionEvent event)
{
// onTouchEvent回调方法重载的屏幕监听方法，该方法根据事件动作的不同执行不同的操作
    switch(event.getAction())
    {
//当前事件为屏幕被按下的事件，通过调用MotionEvent的getX和getY方法得到事件发生的坐标，
//然后设置给自定义View的x与y成员变量
        case MotionEvent.ACTION_DOWN:   //按下
            myView.x = (int) event.getX();       //改变x坐标
            myView.y = (int) event.getY()-52;    //改变y坐标
            myView.postInvalidate();             //重绘
            break;
// 表示在屏幕上滑动时的事件，同样是得到事件发生的位置并设置给View的x、y。需要注意的
//是，因为此时手机屏幕并不是全屏模式，所以需要对坐标进行调整
        case MotionEvent.ACTION_MOVE:   //移动
            myView.x = (int) event.getX();       //改变x坐标
            myView.y = (int) event.getY()-52;    //改变y坐标
            myView.postInvalidate();             //重绘
            break;
// 以下是屏幕被抬起的事件，此时将View的x、y成员变量设成-100，表示并不需要在屏幕中绘制
//矩形
        case MotionEvent.ACTION_UP:     //抬起
            myView.x = -100;                     //改变x坐标
```

```
            myView.y = -100;                        //改变y坐标
            myView.postInvalidate();                //重绘
            break;
        }
        return super.onTouchEvent(event);
    }
    class MyView extends View{                      //自定义的View
        Paint paint;                                //画笔
        int x = 50;                                 //x坐标
        int y = 50;                                 //y坐标
        int w = 80;                                 //矩形的宽度
   public MyView(Context context)
        {//构造器
        super(context);
        paint = new Paint();                        //初始化画笔
    }
    @Override
    protected void onDraw(Canvas canvas)
        {//绘制方法
        canvas.drawColor(Color.GRAY);               //绘制背景色
        canvas.drawRect(x, y, x+w, y+w, paint);     //根据成员变量绘制矩形
        super.onDraw(canvas);
        }
    }
}
```

自定义的View并不会自动刷新，所以每次改变数据模型时都需要调用postInvalidate方法对屏幕刷新。该案例的最终显示效果如图5-1所示。

图5-1　运行效果图

课后作业

一、填空题

1. Android提供了强大的事件处理机制，它包括两套处理机制：一套是_____，另一套是_____。
2. 对于一个Android应用程序来说，_____是必不可少的，用户与应用程序之间的交互便是通过_____来完成的。
3. 一个线程可以_____另一个线程，同一进程中的多个线程之间可以_____。线程之间相互制约，在运行中呈现_____。
4. 线程有_____、_____和_____3种基本状态。

二、选择题

1. （　　）不是多线程机制的特点。
   A. 避免ANR，提升用户体验　　　　B. 实现异步处理
   C. 实现多任务　　　　　　　　　　D. 计算量大
2. （　　）接口的功能为：在触控模式下，是指在某个View上按下并抬起的组合动作；而在键盘模式下，是某个View获得焦点后点击确定键或者按下轨迹球的事件。
   A. OnClickListener　　　　　　　B. OnFocusChangeListener
   C. OnKeyListener　　　　　　　　D. OnCreateContextMenuListener
3. 回调方法（　　）的功能为捕捉手机键盘按键抬起的事件。它是接口KeyEvent.Callback中的一个抽象方法，并且所有的View都实现了该接口并重载了该方法。
   A. onTouchEvent　　　　　　　　B. onTrackBallEvent
   C. onKeyUp　　　　　　　　　　D. onTrackBallEvent
4. 不属于多线程实现过程的是（　　）。
   A. 线程定义　　　　　　　　　　B. 定义子类
   C. Handler、Message和Looper　　D. 线程间的消息传递

三、操作题

利用Android Studio开发环境创建一个新的Android应用程序项目，使用回调方法实现：在用户点击的位置绘制一个矩形，然后监测用户触控笔的状态。当用户在屏幕上移动触控笔时，矩形随之移动；当用户触控笔离开手机屏幕时，停止绘制矩形。（提示：参照书中5.2节内容进行练习。）

第 6 章

数据存储机制

---- 内容概要 ----

本章主要介绍Android平台数据存储的基础知识，以及学习如何在Android应用程序中存储数据。在大多数应用软件开发中，存储数据都是一个非常重要的问题。对于Android应用程序来说，有3种存储数据的基本方式：第1种是轻量级的机制，仅保存少量数据，即Shared Preferences；第2种是通过传统的文件系统API存储数据到文件中；第3种是使用关系型数据库，特别是使用SQLite数据库存取数据。本章主要讲解如何创建和访问应用程序自身的私有数据。如果需要和其他应用程序进行数据共享，则需要用到ContentProvider。

---- 数字资源 ----

【本章案例文件】："案例文件\第6章"目录下

◉ 配套资源
◉ 入门精讲
◉ 项目实战
◉ 日志记录

6.1 Shared Preferences

Shared Preferences是Android系统存储数据的一种方式。Android平台提供SharedPreferences对象来保存简单的应用程序数据，其作用类似于Windows平台上常见的ini文件（用于保存应用程序的一些配置信息）。SharedPreferences对象是通过名值对方式来保存配置信息的，然后把这些名值对写入一个XML文件中，使用起来相当方便。例如，应用程序可能有一个选项允许用户指定应用中显示文本的字体和大小，此时应用程序必须记住用户设置的字体和大小，以便下次再次使用该应用程序时，该应用能够准确地把字体设置为用户上次设置的字体和大小。要达到这种效果，也可以把配置信息保存到文件或数据库中。但是，如果需要保存的配置信息太多，如文本大小、字体、背景色等，则保存到文件将变得非常麻烦。虽然把配置信息保存到数据库中也可以，但是把过多简单数据保存在数据库中会影响应用程序的性能。

下面通过一个案例学习使用SharedPreferences对象存储和修改配置数据。

【案例6-1】：使用SharedPreferences对象保存和修改配置信息，使用PreferenceActivity显示配置信息。

操作步骤如下。

步骤01 创建一个新的Android项目，命名为"LearningPreferences"。

步骤02 在res文件夹中创建一个新的子文件夹并命名为"xml"。在新创建的"xml"文件夹中添加一个新的文件"myappprefs.xml"，该文件的代码如下。

```xml
<?xml version="1.0" encoding="utf-8"?>
<PreferenceScreen
  xmlns:android="http://schemas.android.com/apk/res/android">
  <PreferenceCategory android:title="Category 1">
    <CheckBoxPreference
      android:title="Checkbox"
      android:defaultValue="false"
      android:summary="True or False"
      android:key="checkboxPref" />
  </PreferenceCategory>
  <PreferenceCategory android:title="Category 2">
    <EditTextPreference
      android:summary="Enter a string"
      android:defaultValue="[Enter a string here]"
      android:title="Edit Text"
      android:key="editTextPref"
      />
  </PreferenceCategory>
</PreferenceScreen>
```

步骤 03 创建一个新的类文件，命名为"AppPrefActivity"。打开"AppPrefActivity.java"文件，修改并添加如下代码。

```java
package com.selfteaching.learningpreferences;
import android.preference.PreferenceActivity;
import android.os.Bundle;

public class AppPrefActivity extends PreferenceActivity {
    @Override
    public void onCreate(Bundle savedInstanceState){
        super.onCreate(savedInstanceState);
        //load preferences from the XML file
        addPreferencesFromResource(R.xml.myappprefs);
    }
}
```

步骤 04 在"AndroidManifest.xml"文件中添加AppPrefActivity类的入口，如下面代码中粗体代码所示。

```xml
<?xml version="1.0" encoding="utf-8"?>
<manifest xmlns:android="http://schemas.android.com/apk/res/android"
    package="com.selfteaching.learningpreferences"
    android:versionCode="1"
    android:versionName="1.0" >
    <uses-sdk android:minSdkVersion="14" />

    <application android:icon="@drawable/ic_launcher" android:label="@string/app_name" >
        <activity android:label="@string/app_name" android:name=".LearningPreferencesActivity" >
            <intent-filter >
                <action android:name="android.intent.action.MAIN" />
                <category android:name="android.intent.category.LAUNCHER" />
            </intent-filter>
        </activity>
        <activity android:name=".AppPrefActivity"    android:label="@string/app_name">
            <intent-filter>
                <action android:name="com.selfteaching.AppPrefActivity" />
                <category android:name="android.intent.category.DEFAULT" />
            </intent-filter>
        </activity>
```

```xml
    </application>
</manifest>
```

步骤 05 修改 "main.xml" 文件，进行页面布局，代码如下。

```xml
<?xml version="1.0" encoding="utf-8"?>
<LinearLayout xmlns:android="http://schemas.android.com/apk/res/android"
    android:layout_width="fill_parent"
    android:layout_height="fill_parent"
    android:orientation="vertical" >

<Button
    android:id="@+id/btnPreferences"
    android:text="Load Preferences Screen"
    android:layout_width="fill_parent"
    android:layout_height="wrap_content"
    android:onClick="onClickLoad"/>

<Button
    android:id="@+id/btnDisplayValues"
    android:text="Display Preferences Values"
    android:layout_width="fill_parent"
    android:layout_height="wrap_content"
    android:onClick="onClickDisplay"/>

<EditText
    android:id="@+id/txtString"
    android:layout_width="fill_parent"
    android:layout_height="wrap_content" />

<Button
    android:id="@+id/btnModifyValues"
    android:text="Modify Preferences Values"
    android:layout_width="fill_parent"
    android:layout_height="wrap_content"
    android:onClick="onClickModify"/>
</LinearLayout>
```

步骤 06 修改 "LearningPreferencesActivity.java" 文件，代码如下。

```java
package com.selfteaching.learningpreferences;
import android.app.Activity;
import android.content.Intent;
import android.content.SharedPreferences;
import android.os.Bundle;
import android.view.View;
import android.widget.EditText;
import android.widget.Toast;

public class LearningPreferencesActivity extends Activity {
    @Override
    public void onCreate(Bundle savedInstanceState) {
        super.onCreate(savedInstanceState);
        setContentView(R.layout.main);
    }

    public void onClickLoad(View view) {
        Intent i = new Intent("com.selfteaching.AppPrefActivity");
        startActivity(i);
    }

    public void onClickDisplay(View view) {
        SharedPreferences appPrefs =
getSharedPreferences("com.selfteaching.learningpreferences_preferences", MODE_PRIVATE);

        DisplayText(appPrefs.getString("editTextPref", ""));
    }

    public void onClickModify(View view) {
        SharedPreferences appPrefs =
getSharedPreferences("com.selfteaching.learningpreferences_preferences", MODE_PRIVATE);
        SharedPreferences.Editor prefsEditor = appPrefs.edit();
        prefsEditor.putString("editTextPref",
            ((EditText) findViewById(R.id.txtString)).getText().toString());
        prefsEditor.commit();
```

```
    }

    private void DisplayText(String str) {
        Toast.makeText(getBaseContext(), str, Toast.LENGTH_LONG).show();
    }
}
```

步骤 07 在模拟器中运行该应用程序。点击"Load Preferences Screen"按钮，显示Preferences界面，如图6-1所示。

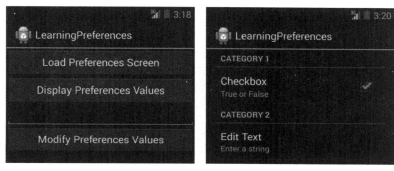

图 6-1 显示 Preferences 界面

步骤 08 点击Perferences界面上的Checkbox（复选框）或在Edit Text（文本编辑项）中输入文本信息，使原始Preferences的值发生变化，然后离开Preferences界面。之后，一个新文件将会在"\data\data\com.selfteaching.learningpreferences\shared_prefs"文件夹中被创建出来，如图6-2所示。为了查看该文件，切换到Eclipse的DDMS视图，打开"File Explorer"选项卡，将会看见一个新的XML文件，名为"com.selfteaching.learningpreferences_preferences.xml"。Preferences界面中原始Preferences值的变化都保存在该文件中。

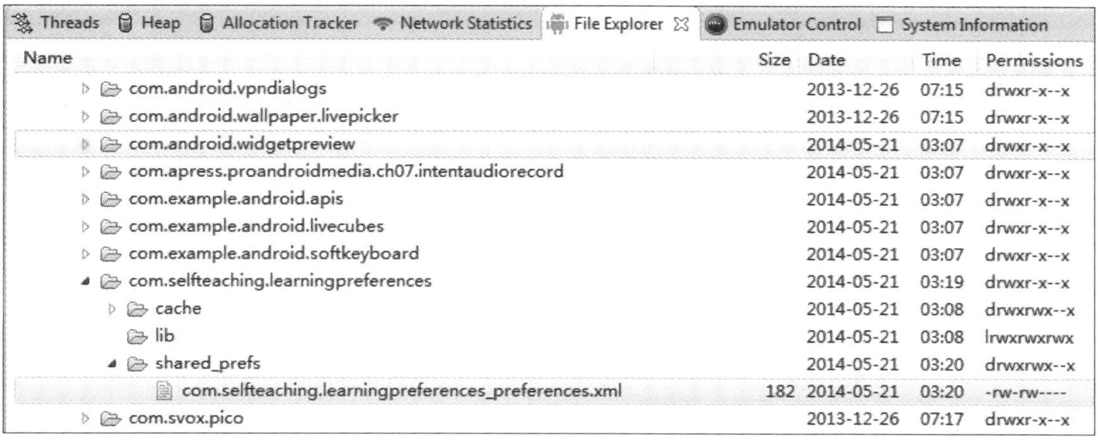

图 6-2 "preferences.xml" 文件

步骤09 打开"preferences.xml"文件，查看该文件的内容，代码如下。

```xml
<?xml version="1.0" encoding="utf-8" standalone="yes" ?>
<map>
<string name="editTextPref">self-teaching</string>
<boolean name="checkboxPref" value="true" />
</map>
```

步骤10 点击"Display Preferences Values"按钮，显示如图6-3所示的界面，然后在EditText中输入文本内容并且点击"Modify Preferences Values"按钮，如图6-4所示。

步骤11 再次点击"Display Preferences Values"按钮，发现新修改的值被保存下来，如图6-5所示。

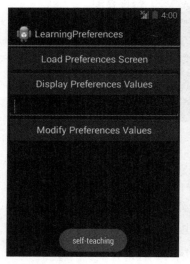 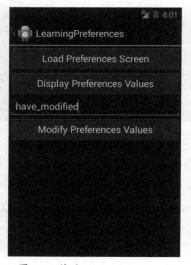 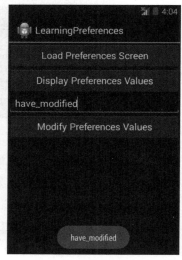

图 6-3　显示 Preferences Values　　　图 6-4　修改 Preferences Values　　　图 6-5　显示修改后的值

代码说明：

（1）本例步骤02中创建的名为"myappprefs.xml"的XML文件用于指定应用程序要保存的preferences类型。代码中创建了两个preferences类别，用于preferences类型的分组，第1个类别中包含一个CheckBoxPreference，其key值为"checkboxPref"；第2个类别中包含一个EditText-Preference，其key值为"editTextPref"。android:key属性用于指定在代码中可以引用的key值，用它来设置或获取相应preference的值。

（2）本例步骤03中创建了一个扩展自基类PreferenceActivity的派生类"AppPrefActivity"。为了便于用户编辑和显示这些preferences，需要调用addPreferencesFromResource方法以加载包含preferences的XML文件，参见步骤03中的代码。

（3）在步骤06中通过使用一个Intent对象，启动Activity，关键的代码是以下这两行。

```
Intent i = new Intent("com.selfteaching.AppPrefActivity");
startActivity(i);
```

（4）preferences的所有改变将会自动存储到应用程序的"shared_prefs"目录下的一个XML文件中。为了显示preferences中的配置信息，在步骤06的onClickDisplay方法中，首先使用getSharedPreferences方法获取SharedPreferences对象。本例需要把上面自动创建的XML文件的文件名传给getSharedPreferences方法，该XML文件的命名方式是<PackageName>_preferences。为了获取某个具体的preference的值，需使用getString方法把某个preference的key值传给该方法。为方便阅读，这里将步骤06中的onClickDisplay方法单独列出。

```java
public void onClickDisplay(View view) {
    SharedPreferences appPrefs =
    getSharedPreferences("com.selfteaching.learningpreferences_preferences", MODE_PRIVATE);
    DisplayText(appPrefs.getString("editTextPref", ""));
}
```

代码中MODE_PRIVATE常量表明该preference文件只能由创建它的应用程序打开。

（5）在步骤06的onClickModify方法中，首先需要创建SharedPreferences.Editor对象，该对象由SharedPreferences对象的edit方法创建。为了更改字符串的preference值，需使用putString方法。为了将改变的值保存到XML文件中，需调用commit方法。为方便阅读，这里将步骤06中的onClickModify方法单独列出。

```java
public void onClickModify(View view) {
    SharedPreferences appPrefs =
    getSharedPreferences("com.selfteaching.learningpreferences_preferences", MODE_PRIVATE);
    SharedPreferences.Editor prefsEditor = appPrefs.edit();
    prefsEditor.putString("editTextPref",
        ((EditText) findViewById(R.id.txtString)).getText().toString());
    prefsEditor.commit();
}
```

6.2 存储数据到文件

SharedPreferences对象允许存储名值对数据，但有时也需要使用文件系统来存储数据。在Android平台上，可以使用java.io包中的类将数据存储到文件中。本节将介绍如何存储数据到内容存储器和外部存储卡中。

6.2.1 实现过程

在Android应用程序中，保存文件的第1种方式就是把数据写到设备的内部存储中。

【案例6-2】：把用户输入的文本内容保存到设备的内部存储中。

操作步骤如下。

步骤 01 创建一个新的Android项目，命名为"LearningFiles"。

步骤 02 修改"main.xml"文件，代码如下。

```xml
<?xml version="1.0" encoding="utf-8"?>
<LinearLayout xmlns:android="http://schemas.android.com/apk/res/android"
    android:layout_width="fill_parent"
    android:layout_height="fill_parent"
    android:orientation="vertical" >

<TextView
    android:layout_width="fill_parent"
    android:layout_height="wrap_content"
    android:text="Please enter some text" />

<EditText
    android:id="@+id/txtText1"
    android:layout_width="fill_parent"
    android:layout_height="wrap_content" />

<Button
    android:id="@+id/btnSave"
    android:text="Save"
    android:layout_width="fill_parent"
    android:layout_height="wrap_content"
    android:onClick="onClickSave" />

<Button
    android:id="@+id/btnLoad"
    android:text="Load"
    android:layout_width="fill_parent"
    android:layout_height="wrap_content"
    android:onClick="onClickLoad" />

</LinearLayout>
```

步骤 03 修改"LearningFilesActivity.java"文件，代码如下。

```java
package com.selfteaching.learningfiles;
import java.io.BufferedReader;
import java.io.File;
import java.io.FileInputStream;
import java.io.FileOutputStream;
import java.io.IOException;
import java.io.InputStream;
import java.io.InputStreamReader;
import java.io.OutputStreamWriter;
import android.app.Activity;
import android.os.Bundle;
import android.os.Environment;
import android.view.View;
import android.widget.EditText;
import android.widget.Toast;

public class LearningFilesActivity extends Activity {
    EditText textBox;
    static final int READ_BLOCK_SIZE = 100;

    @Override
    public void onCreate(Bundle savedInstanceState) {
        super.onCreate(savedInstanceState);
        setContentView(R.layout.main);
        textBox = (EditText) findViewById(R.id.txtText1);
    }

    public void onClickSave(View view) {
        String str = textBox.getText().toString();
        try
        {
            FileOutputStream fOut =openFileOutput("textfile.txt", MODE_WORLD_READABLE);
            OutputStreamWriter osw = new OutputStreamWriter(fOut);

            //---write the string to the file---
            osw.write(str);
```

```
        osw.flush();
        osw.close();

        //---display file saved message---
        Toast.makeText(getBaseContext(),
           "File saved successfully!", Toast.LENGTH_SHORT).show();

        //---clears the EditText---
        textBox.setText("");
    }
    catch (IOException ioe)
    {
        ioe.printStackTrace();
    }
}

public void onClickLoad(View view) {
    try
    {
        FileInputStream fIn = openFileInput("textfile.txt");
        InputStreamReader isr = new InputStreamReader(fIn);

        char[] inputBuffer = new char[READ_BLOCK_SIZE];
        String s = "";

        int charRead;
        while ((charRead = isr.read(inputBuffer))>0)
        {
            //---convert the chars to a String---
            String readString = String.copyValueOf(inputBuffer, 0, charRead);
            s += readString;

            inputBuffer = new char[READ_BLOCK_SIZE];
        }
        //---set the EditText to the text that has been read---
        textBox.setText(s);
        Toast.makeText(getBaseContext(),
```

```
            "File loaded successfully!", Toast.LENGTH_SHORT).show();
        }
        catch (IOException ioe) {
            ioe.printStackTrace();
        }
    }
}
```

步骤 04 在模拟器中运行该应用程序。在文本编辑框中输入文本"hello files",然后点击"Save"按钮,如图6-6所示。

图 6-6 输入文本

步骤 05 文件保存成功会显示文件保存成功的提示信息,然后文本框中的文本内容将会消失,如图6-7所示。

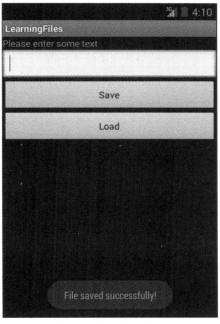

图 6-7 文件保存成功

步骤06 点击"Load"按钮,当"hello files"字符串再次出现在文本框时,说明文件被成功保存了,如图6-8所示。

图 6-8 加载文件

■6.2.2 操作分析

保存文本到文件中,需要使用FileOutputStream类,该类中的openFileOutput方法用于打开一个文件。代码中使用的MODE_WORLD_READABLE常量表示该文件对所有应用来说都是可读的。步骤03中的代码如下。

```
FileOutputStream fOut = openFileOutput("textfile.txt", MODE_WORLD_READABLE);
```

除了MODE_WORLD_READABLE常量之外,还可选择常量MODE_PRIVATE,表明该文件仅能由创建它的应用访问;选择常量MODE_APPEND,表明向文件中追加内容;选择常量MODE_WORLD_WRITEABLE,表明所有应用都可写入该文件。

创建OutputStreamWriter类的一个实例,把已创建的FileOutputStream类对象"fOut"传给它,实现字符流到字节流的转换。实现此功能的代码如下。

```
OutputStreamWriter osw = new OutputStreamWriter(fOut);
```

使用write方法将字符串写入文件;为了确保所有字节都写入文件,需要使用flush方法刷新;最后,再使用close方法关闭文件。实现这些功能的代码如下。

```
//---write the string to the file---
osw.write(str);
```

```
osw.flush();
osw.close();
```

为了读出文件内容，需要使用FileInputStream类和InputStreamReader类。实现此功能的两行代码如下。

```
FileInputStream fIn = openFileInput("textfile.txt");
InputStreamReader isr = new InputStreamReader(fIn);
```

因为事先不知道文件的大小，于是先把内容读到一个100字符的缓冲块中，然后把读取的字符块复制到一个String对象中。实现此功能的是以下这段代码。

```
char[] inputBuffer = new char[READ_BLOCK_SIZE];
String s = "";

int charRead;
while ((charRead = isr.read(inputBuffer))>0)
{
  //---convert the chars to a String---
  String readString = String.copyValueOf(inputBuffer, 0, charRead);
  s += readString;

  inputBuffer = new char[READ_BLOCK_SIZE];
}
```

在模拟器中运行应用程序，切换到DDMS视图，查看应用是否创建了一个新文件，文件路径是"\data\data\com.selfteaching.learningfiles\files"，如图6-9所示。

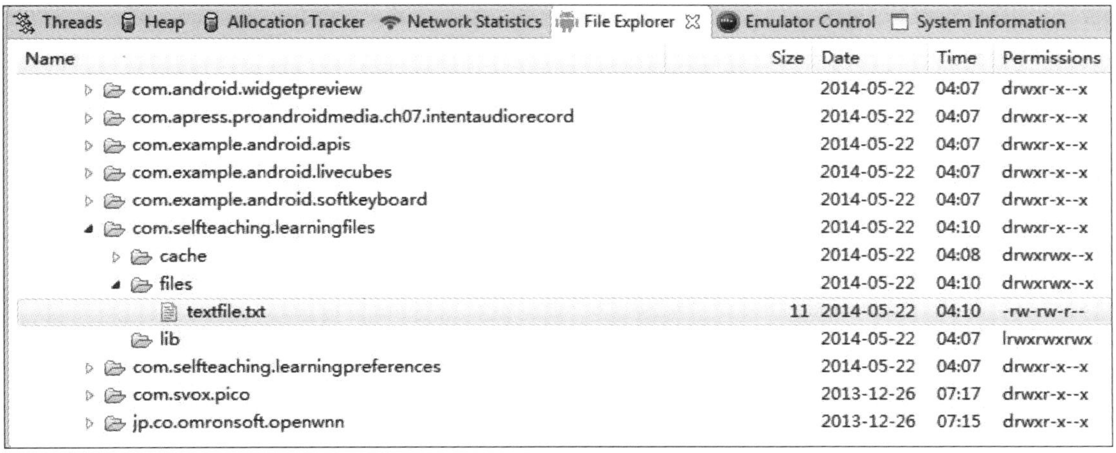

图 6-9　DDMS 视图中查看新创建的文件

以上介绍的是实现数据的内部存储,接着将介绍如何实现将数据保存到外部存储卡中。外部存储卡通常有更大的容量,也更容易共享数据,因此将数据保存到外部存储卡上是一个更好的选择。

操作步骤如下。

步骤 01 使用6.2.1节中的项目,修改"LearningFilesActivity.java"文件,代码如下。

```java
public void onClickSave(View view) {
    String str = textBox.getText().toString();
    try
    {
        //---SD Card Storage---
        File sdCard = Environment.getExternalStorageDirectory();
        File directory = new File (sdCard.getAbsolutePath() + "/myfiles");
        directory.mkdirs();
        File file = new File(directory, "textfile.txt");
        FileOutputStream fOut = new FileOutputStream(file);

        OutputStreamWriter osw = new OutputStreamWriter(fOut);
        //---write the string to the file---
        osw.write(str);
        osw.flush();
        osw.close();
        //---display file saved message---
        Toast.makeText(getBaseContext(), "File saved successfully!",
            Toast.LENGTH_SHORT).show();
        //---clears the EditText---
        textBox.setText("");
    }catch (IOException ioe){
        ioe.printStackTrace();
    }
}

public void onClickLoad(View view) {
    try
    {
        //---SD Storage---
        File sdCard = Environment.getExternalStorageDirectory();
        File directory = new File (sdCard.getAbsolutePath() + "/myfiles");
```

```java
        File file = new File(directory, "textfile.txt");
        FileInputStream fIn = new FileInputStream(file);
        InputStreamReader isr = new InputStreamReader(fIn);

        char[] inputBuffer = new char[READ_BLOCK_SIZE];
        String s = "";
        int charRead;
        while ((charRead = isr.read(inputBuffer))>0)
        {
            //---convert the chars to a String---
            String readString = String.copyValueOf(inputBuffer, 0, charRead);
            s += readString;
            inputBuffer = new char[READ_BLOCK_SIZE];
        }
        ... ...
}
```

步骤 02 修改 "AndroidManifest.xml" 文件，代码如下。

```xml
<?xml version="1.0" encoding="utf-8"?>
<manifest xmlns:android="http://schemas.android.com/apk/res/android"
    package="com.selfteaching.learningfiles"
    android:versionCode="1"
    android:versionName="1.0" >
    <uses-sdk android:minSdkVersion="14" />
    <uses-permission android:name="android.permission.WRITE_EXTERNAL_STORAGE" />
    <application
        android:icon="@drawable/ic_launcher"
        android:label="@string/app_name" >
        <activity
            android:label="@string/app_name"
            android:name=".LearningFilesActivity" >
            <intent-filter >
                <action android:name="android.intent.action.MAIN" />
                <category android:name="android.intent.category.LAUNCHER" />
            </intent-filter>
        </activity>
    </application>
</manifest>
```

在上述操作过程中，getExternalStorageDirectory方法可以返回外部存储卡的全路径。通常在真实设备上，该方法会返回到"\sdcard"，而在模拟器上会返回到"\mnt\sdcard"，不要硬性设定存储卡的路径，因为设备商可能会更改存储卡的路径，所以最好始终使用getExternalStorageDirectory方法去获取外部存储卡的全路径。

最后，使用File对象的mkdirs方法在存储卡上创建一个目录用于保存文件，实现此功能的代码如下：

```
File directory = new File (sdCard.getAbsolutePath() + "/myfiles");
directory.mkdirs();
```

为了能够写数据到外部存储卡中，需要有相应的写权限，因此，需要在"AndroidManifest.xml"文件中添加WRITE_EXTERNAL_STORAGE权限。

在模拟器上运行应用程序，可以在"\mnt\sdcard\myfiles"文件夹下看到新创建的文件。

6.3 使用数据库存储数据

到目前为止，前面介绍的技术只能存储简单的数据。如需保存关系型数据，使用数据库会更加高效。例如，如果要保存学校里所有学生的测验成绩，使用数据库会更加高效。因为使用数据库不仅能更方便地检索学生的成绩，而且使用数据库保存数据能更好地保证数据完整性。Android平台使用SQLite数据库系统。为一个应用程序创建的数据库只能被它自己访问，其他应用程序没有访问权限。在Android平台中，应用程序创建的SQLite数据库总是被存放在"\data\data\<package_name>\databases"文件夹下。本节主要介绍如何在应用程序中动态地编写创建、增加、删除、修改、查询数据库的程序。

创建一个数据库帮助类DBAdapter来打包复杂的数据库操作是一个良好的编码习惯，它可以隐藏具体的数据库操作细节，使用户只需关注业务逻辑的实现。

6.3.1 创建数据库帮助类DBAdapter

【案例6-3】：创建一个名为"myfirstdb"的数据库，其中包含一张名为"books"的数据表。该表有3列，分别是_id、name和authors。通过该案例介绍如何创建、打开、关闭及操作数据库。

操作步骤如下。

步骤01 创建一个新的Android项目，命名为"LearningDatabases"。

步骤02 在src文件夹中添加一个新的Java源代码文件，命名为"DBAdapter"，该Java源代码文件需要包含以下几部分。

（1）常量定义部分。

在编写的DBAdapter类代码中，先定义几个常量，分别表示数据表的字段名、数据库名、数据表名和创建数据表"books"的SQL语句。

```
static final String KEY_ROWID = "_id";
static final String KEY_NAME = "name";
static final String KEY_AUTHORS = "authors";
static final String TAG = "DBAdapter";

static final String DATABASE_NAME = "myfirstdb";
static final String DATABASE_TABLE = "books";
static final int DATABASE_VERSION = 2;

static final String DATABASE_CREATE =
    "create table books (_id integer primary key autoincrement, "
    + "name text not null, authors text not null);";
```

（2）在DBAdapter类中，创建了一个扩展自SQLiteOpenHelper类的私有类"DatabaseHelper"，负责管理数据库的创建和版本管理。在该私有类中重载了onCreate方法和onUpgrade方法，具体代码如下。

```
private static class DatabaseHelper extends SQLiteOpenHelper
{
    DatabaseHelper(Context context)
    {
        super(context, DATABASE_NAME, null, DATABASE_VERSION);
    }

    @Override
    public void onCreate(SQLiteDatabase db)
    {
      try {
        db.execSQL(DATABASE_CREATE);
      } catch (SQLException e) {
        e.printStackTrace();
      }
    }

    @Override
    public void onUpgrade(SQLiteDatabase db, int oldVersion, int newVersion)
    {
```

```
        Log.w(TAG, "Upgrading database from version " + oldVersion + " to "
            + newVersion + ", which will destroy all old data");
        db.execSQL("DROP TABLE IF EXISTS books");
        onCreate(db);
    }
}
```

如果数据库不存在，先用onCreate方法创建一个新的数据库"myfirstdb"。当数据库需要更新时，会调用onUpgrade方法。通过检测在DATABASE_VERSION常量中的值来确定是否需要更新数据库。onUpgrade方法的具体实现是简单地删除数据表"books"，然后再重新创建它。

（3）定义打开和关闭数据库的方法，以及增加、删除和修改数据表中记录的方法。具体代码如下。

```
//---opens the database---
    public DBAdapter open() throws SQLException
    {
        db = DBHelper.getWritableDatabase();
        return this;
    }

//---closes the database---
    public void close()
    {
        DBHelper.close();
    }

//---insert a book into the database---
    public long insertBook(String name, String authors)
    {
        ContentValues initialValues = new ContentValues();
        initialValues.put(KEY_NAME, name);
        initialValues.put(KEY_AUTHORS, authors);
        return db.insert(DATABASE_TABLE, null, initialValues);
    }

//---deletes a particular book---
    public boolean deleteBook(long rowId)
    {
```

```
    return db.delete(DATABASE_TABLE, KEY_ROWID + "=" + rowId, null) > 0;
}


//---retrieves all the books---
public Cursor getAllBooks()
{
    return db.query(DATABASE_TABLE, new String[] {KEY_ROWID, KEY_NAME,
        KEY_AUTHORS}, null, null, null, null, null);
}


//---retrieves a particular book---
public Cursor getBook(long rowId) throws SQLException
{
    Cursor mCursor =
        db.query(true, DATABASE_TABLE, new String[] {KEY_ROWID,
        KEY_NAME, KEY_AUTHORS}, KEY_ROWID + "=" + rowId, null,
        null, null, null, null);
    if (mCursor != null) {
        mCursor.moveToFirst();
    }
    return mCursor;
}


//---updates a book---
public boolean updateBook(long rowId, String name, String authors)
{
    ContentValues args = new ContentValues();
    args.put(KEY_NAME, name);
    args.put(KEY_AUTHORS, authors);
    return db.update(DATABASE_TABLE, args, KEY_ROWID + "=" + rowId, null) > 0;
}
```

　　Android使用Cursor类作为数据库查询的返回值。可以把Cursor看成指向数据库查询出的结果集的指针。使用Cursor可使Android更加高效地管理记录和字段。
　　ContentValues对象可以存储名值对，它的put方法用于插入具有不同数据类型值的名值对。
　　使用DBAdapter类在应用程序中创建数据库时，首先要创建DBAdapter类的实例。创建实例就需要用类的构造函数，DBAdapter类构造函数的具体代码如下。

```java
public DBAdapter(Context ctx)
{
    this.context = ctx;
    DBHelper = new DatabaseHelper(context);
}
```

DBAdapter类的构造函数中需要创建DatabaseHelper类的实例,即调用DatabaseHelper类的构造函数创建该类的一个实例。DatabaseHelper类的构造函数的具体代码如下。

```java
public DatabaseHelper(Context context)
{
    super(context, DATABASE_NAME, null, DATABASE_VERSION);
}
```

6.3.2 添加数据到数据表books中

使用6.3.1节中的项目,修改"LearningDatabasesActivity.java"文件,代码如下。

```java
package com.selfteaching.learningdatabases;
import android.app.Activity;
import android.os.Bundle;

public class LearningDatabasesActivity extends Activity {
    @Override
    public void onCreate(Bundle savedInstanceState) {
        super.onCreate(savedInstanceState);
        setContentView(R.layout.main);

        DBAdapter db = new DBAdapter(this);
        //---add two book---
        db.open();
        long id = db.insertBook("Learning Android", "Jim");
        id = db.insertBook("Learning Python", "Tom");
        db.close();
    }
}
```

本例代码中先创建一个DBAdapter类的实例,然后调用它的insertBook()方法插入两条记录,即向数据表"books"中插入两本书的记录,并返回插入记录的id值。如果插入时有错误发生,则返回-1。

切换到DDMS视图,查看File Explorer选项卡,会发现此时在"databases"文件夹下出现了一个名为"myfirstdb"的文件,说明"myfirstdb"数据库文件已经建立。

6.3.3 获取数据表中的所有记录

仍然使用6.3.1节中的项目,修改"LearningDatabasesActivity.java"文件,代码如下。

```
package com.selfteaching.learningdatabases;
import android.app.Activity;
import android.os.Bundle;

public class LearningDatabasesActivity extends Activity {
    @Override
    public void onCreate(Bundle savedInstanceState) {
        super.onCreate(savedInstanceState);
        setContentView(R.layout.main);

        DBAdapter db = new DBAdapter(this);

        /*
        //---add two books---
        db.open();
        long id = db.insertBook("Learning Android", "Jim");
        id = db.insertBook("Learning Python", "Tom");
        db.close();
        */

        //--get all contacts---
        db.open();
        Cursor c = db.getAllContacts();
        if (c.moveToFirst())
        {
            do {
                DisplayBook(c);
            } while (c.moveToNext());
        }
```

```
        db.close();
    }
    public void DisplayBook(Cursor c)
    {
        Toast.makeText(this,
            "id: " + c.getString(0) + "\n" +
            "Name: " + c.getString(1) + "\n" +
            "Authors: " + c.getString(2),
            Toast.LENGTH_LONG).show();
    }
}
```

在以上代码中，用DBAdapter类的getAllBooks方法获取数据库中的所有书的记录，返回结果是一个Cursor对象。为了显示所有书的记录，必须首先调用Cursor对象的moveToFirst方法，目的是将指针移动到第1条记录。如果成功，表明返回结果至少有一条记录，然后调用DisplayBook方法显示书的详情。之后再调用Cursor对象的moveToNext方法将指针移到下一条记录的位置。

6.3.4 获取数据表中的某一条记录

使用6.3.3的项目，修改"LearningDatabasesActivity.java"文件，代码如下。

```
package com.selfteaching.learningdatabases;
import android.app.Activity;
import android.os.Bundle;

public class LearningDatabasesActivity extends Activity {

    @Override
    public void onCreate(Bundle savedInstanceState) {
        super.onCreate(savedInstanceState);
        setContentView(R.layout.main);

        DBAdapter db = new DBAdapter(this);

        /*
        //---add two books---
        db.open();
```

```
       long id = db.insertBook("Learning Android", "Jim");
       id = db.insertBook("Learning Python", "Tom");
       db.close();
       */

       /*
       //--get all books---
       db.open();
       Cursor c = db.getAllContacts();
       if (c.moveToFirst())
       {
          do {
             DisplayBook(c);
          } while (c.moveToNext());
       }
       db.close();
       */

       //---get a book---
       db.open();
       Cursor c = db.getBook(2);
       if (c.moveToFirst())
          DisplayBook(c);
       else
          Toast.makeText(this, "No book found", Toast.LENGTH_LONG).show();
       db.close();
    }

    public void DisplayBook(Cursor c)
    {
       Toast.makeText(this,
             "id: " + c.getString(0) + "\n" +
             "Name: " + c.getString(1) + "\n" +
             "Authors:  " + c.getString(2),
             Toast.LENGTH_LONG).show();
    }
}
```

在模拟器中运行该应用程序，运行结果是第2条记录的详细信息显示在界面上。

在上面的代码中，用DBAdapter类的getBook方法返回由参数id指定的单条记录。本例中传给getBook方法的参数id的值为2，表示要获取的是第2条记录，即如下这行代码。

```
Cursor c = db.getBook(2);
```

这行代码的返回结果是一个Cursor对象。如果一条记录被返回，还需要调用DisplayBook方法才能显示该记录的详情，否则不会显示任何信息。

■6.3.5 更新数据表中的某一条记录

使用6.3.4节的项目，继续修改"LearningDatabasesActivity.java"文件，代码如下。

```java
package com.selfteaching.learningdatabases;
import android.app.Activity;
import android.os.Bundle;

public class LearningDatabasesActivity extends Activity {

    @Override
    public void onCreate(Bundle savedInstanceState) {
        super.onCreate(savedInstanceState);
        setContentView(R.layout.main);

        DBAdapter db = new DBAdapter(this);

        /*
        //---add two books---
        db.open();
        long id = db.insertBook("Learning Android", "Jim");
        id = db.insertBook("Learning Python", "Tom");
        db.close();
        */

        /*
        //--get all books---
        db.open();
        Cursor c = db.getAllContacts();
        if (c.moveToFirst())
```

```
      {
        do {
          DisplayBook(c);
        } while (c.moveToNext());
      }
      db.close();
    */


    /*
    //---get a book---
    db.open();
    Cursor c = db.getBook(2);
    if (c.moveToFirst())
        DisplayBook(c);
    else
        Toast.makeText(this, "No book found", Toast.LENGTH_LONG).show();
    db.close();
    */


    //---update contact---
    db.open();
    if (db.updateBook(1, "Learning Android 2", "Smith"))
        Toast.makeText(this, "Update successful.", Toast.LENGTH_LONG).show();
    else
        Toast.makeText(this, "Update failed.", Toast.LENGTH_LONG).show();
    db.close();
}


public void DisplayBook(Cursor c)
{
    Toast.makeText(this,
        "id: " + c.getString(0) + "\n" +
        "Name: " + c.getString(1) + "\n" +
        "Authors:  " + c.getString(2),
        Toast.LENGTH_LONG).show();
}
}
```

在上面的代码中，用DBAdapter类的updateBook方法更新记录的详情，设置的参数是需要更新的某本书的id值、书名、作者3个参数。该方法返回的是布尔值，布尔值表明是否更新成功。

最后，在模拟器中运行该应用程序，界面中会显示更新是否成功的提示信息。

6.3.6 删除数据表中的某一条记录

使用6.3.5节的项目，继续修改"LearningDatabasesActivity.java"文件，代码如下。

```java
package com.selfteaching.learningdatabases;
import android.app.Activity;
import android.os.Bundle;

public class LearningDatabasesActivity extends Activity {

    @Override
    public void onCreate(Bundle savedInstanceState) {
        super.onCreate(savedInstanceState);
        setContentView(R.layout.main);

        DBAdapter db = new DBAdapter(this);

        /*
        //---add two books---
        db.open();
        long id = db.insertBook("Learning Android", "Jim");
        id = db.insertBook("Learning Python", "Tom");
        db.close();
        */

        /*
        //--get all books---
        db.open();
        Cursor c = db.getAllContacts();
        if (c.moveToFirst())
        {
            do {
                DisplayBook(c);
            } while (c.moveToNext());
```

        }
        db.close();
        */

        /*
        //---get a book---
        db.open();
        Cursor c = db.getBook(2);
        if (c.moveToFirst())
            DisplayBook(c);
        else
            Toast.makeText(this, "No book found", Toast.LENGTH_LONG).show();
        db.close();
        */

        /*
        //---update contact---
        db.open();
        if (db.updateBook(1, "Learning Android 2", "Smith"))
            Toast.makeText(this, "Update successful.", Toast.LENGTH_LONG).show();
        else
            Toast.makeText(this, "Update failed.", Toast.LENGTH_LONG).show();
        db.close();
        */

        //---delete a contact---
        db.open();
        if (db.deleteBook(1))
            Toast.makeText(this, "Delete successful.", Toast.LENGTH_LONG).show();
        else
            Toast.makeText(this, "Delete failed.", Toast.LENGTH_LONG).show();
        db.close();
    }

    public void DisplayBook(Cursor c)
    {
        Toast.makeText(this,

```
                "id: " + c.getString(0) + "\n" +
                "Name: " + c.getString(1) + "\n" +
                "Authors:  " + c.getString(2),
                Toast.LENGTH_LONG).show();
    }
}
```

在上面的代码中，用DBAdapter类的deleteBook方法删除一条记录，设置的参数是需要删除的记录的id值。该方法返回的是布尔值，布尔值表明是否删除成功。

最后，在模拟器中运行该应用程序，删除后会显示删除是否成功的提示信息。

课后作业

一、填空题

1. Android中应用程序的参数设置、运行状态数据只有保存到_____上，系统在关机之后数据才_____。

2. Android提供的3种数据存储的基本方式为：_____、_____、_____。

3. _____是Android平台上一个轻量级的存储类，是基于XML文件来存储_____数据，通常用来存储一些简单的配置信息。

4. Android系统下的文件可以分为两类：一类是_____，另一类是_____。

5. 在Android资源文件中，有两个特殊的文件夹：_____和_____，用于存放app所需的特殊文件，且该文件打包时不会被编码到二进制文件。

二、选择题

1. 在Android中，以下（   ）不是实现SharedPreferences存储的步骤。

    A. 通过Editor方法提交数据

    B. 利用edit方法获取Editor对象

    C. 通过Editor对象存储key-value名值对数据

    D. 根据Context获取SharedPreferences对象

2. Android中的Preference XML文件中的View是有限的，下面的哪个不是（   ）Preference XML文件中的View。

    A. CheckBoxPreference

    B. EditTextPreference

    C. MainActivity

    D. PreferenceScreen

3.Android提供了一组特有的API来访问私有文件，（　　）不属于操作模式。

　　A. Context.MODE_WORLD_WRITEABLE

　　B. Context.MODE_APPEND

　　C. Context.MODE_WORLD_READABLE

　　D. FileInputStream openFileInput

4.Android中的WRITE_EXTERNAL_STORAGE属于危险权限，除了在"AndroidManifest.xml"文件中添加权限外，还需要在运行时进行动态授权，以下（　　）不是其中的步骤。

　　A. 请求权限

　　B. 执行权限相关代码

　　C. 处理结果

　　D. 检查权限

三、操作题

　　数据是应用程序的核心，也是Android程序开发人员和Android用户关注的重点。掌握使用文件系统存储数据很重要，参照书中6.2节内容，练习将数据写到设备的内部存储中。

第 7 章
Intent 和 ContentProvider

内容概要

Android应用主要由四种组件组成，分别为Activity、Broadcast、Service和ContentProvider，而这些组件（ContentProvider除外）之间的通信主要是由Intent协助完成，它适用于传递数据量小的场合。另外，在系统的多个应用程序之间有时需要进行数据的共享与交换，因而Android提供了ContentProvider机制。本章将详细介绍Intent和ContentProvider的使用。

数字资源

【本章案例文件】："案例文件\第7章"目录下

7.1 Intent

Android提供了Intent机制来协助应用间的交互与通信，它封装了Android应用程序需要启动某个组件的"意图"，不仅可用于应用程序之间，还可用于应用程序内部的Activity/Service之间的交互。通过Intent可以启动另一个Activity、启动Service、发起广播等，并可以通过它传递数据。因此，Intent起着媒体中介的作用，专门提供组件之间互相调用的相关信息，实现调用者与被调用者之间的解耦。

7.1.1 Intent的组成

Intent由组件名称、执行动作Action、与动作关联数据的描述等几部分组成。下面阐述各个部分的含义及其作用。

（1）Component。

Component是指定Intent的目标组件的类名称。如果Component属性已指定，将直接使用它指定的组件。指定了Component属性以后，Intent的其他属性都是可选的。

（2）Action。

Action是指要执行的动作，一般使用一个字符串描述将要执行的动作。为了方便引用，Intent类中定义了一些标准的动作，用户也可以根据需要自定义Action。下面是一些常用的Action。

- ACTION_CALL：拨打Data里用Uri表示的电话号码。
- ACTION_MAIN：启动项目的初始界面。
- ACTION_VIEW：常与特定的数据和Uri配合使用，用于将数据和网站等信息显示给用户。例如：

Uri oneUri=Uri.parse（"http://www.baidu.com"）;//指定Uri为网址
Intent aIntent=new Intent(Intent. ACTION_VIEW, oneUri);
startAcivtiy(aIntent);

- ACTION_DIAL：用于描述给用户打电话的动作。例如：

Intent aIntent=new Intent(Intent. ACTION_DIAL, Uri.parse（"tel:123456"）);
startAcivtiy(aIntent);

- ACTION_EDIT：打开数据里指定数据所对应的编辑程序。
- ACTION_DELETE：删除指定的数据。
- ACTION_BATTERY_LOW：警告电池电量低。
- ACTION_HEADSET_PLUG：插入或拔掉耳机设备。
- ACTION_TIME_CHANGED：改变系统时间。

（3）Data。

Data是指要操作的数据，Android中采用指向数据的一个Uri来表示。Data主要完成对Intent消息中数据的封装，不同类型的Action会有不同的Data封装。例如，ACTION_EDIT指定Data为文件Uri，打电话为tel:Uri，访问网络为http:Uri，而由Content Provider提供的数据则为content: URIs。

（4）Category。

Category是对目标组件类别信息的描述。一个Intent对象可以包含多个Category。Intent类定义了许多Category常数来表示Intent的不同类别。下面是一些常用的Category常数。

- **CATEGORY_DEFAULT**：表示默认的Category。
- **CATEGORY_HOME**：设置该Activity随系统启动而运行。
- **CATEGORY_LAUNCHER**：表示该Activity是应用程序中最先被执行的。
- **CATEGORY_BROWSABLE**：该Activity能被浏览器安全调用。
- **CATEGORY_PREFERENCE**：该Activity是参数面板。

（5）Extra。

Extra中封装了一些额外的以名值对形式存在的附加信息。使用Extras可以为组件提供扩展信息，例如，如果要执行"发送电子邮件"这个动作，可以将电子邮件的标题、正文等保存在Extras里传给电子邮件发送组件。Intent可以通过putExtras方法和getExtras方法来存储和获取Extra。

7.1.2 Intent Filter

Intent过滤器描述了组件的一种能力，即乐意接收的一组intents。实际上，过滤器的目的是筛掉不想要的intents。通常Intent过滤器不在Java代码中设置，而是在应用程序的清单文件"AndroidManifest.xml"中的<intent-filter>元素中设置。但有一个例外，广播接收者的过滤器通过调用Context.registerReceiver()动态注册，它直接创建一个IntentFilter对象。一个Intent过滤器是一个IntentFilter类的实例。因为Android系统在启动一个组件之前必须知道它的能力，所以一个过滤器必须有对应于Intent对象的动作、数据、种类的字段。

（1）检测Action。

Action主要的内容有MAIN、VIEW、PICK、EDIT等。清单文件中的<intent-filter>元素使用<action>子元素列出动作，例如：

```
<intent-filter ...... >
    <action android:name="com.example.project.SHOW_CURRENT" />
    <action android:name="com.example.project.SHOW_RECENT" />
    <action android:name="com.example.project.SHOW_PENDING" />
    ......
</intent-filter>
```

一个Intent过滤器必须至少包含一个<action>子元素，否则它将阻塞所有的intents。一般一个Intent对象只能设置一个Action，如果Intent对象没有指定动作，将自动通过检查。但是，一个<intent-filter>可以设置多个Action过滤，只要一个满足即可完成Action验证。

（2）检测Category。

<intent-filter>同样可以设置多个Category。当Intent中的Category属性与<intent-filter>中的一个category完全匹配时，便可通过Category检测，但其他的Category并不受影响。当<intent-filter>中没有设置Category时，只能与没有设置Category的Intent相匹配，原则上应该都能通过种类测试，而不管过滤器中有什么种类。有一种情况例外，Android对于所有传递给Context.startActivity()的隐式Intent应该至少包含"android.intent.category.DEFAULT"（对应CATEGORY_DEFAULT常量）。因此，想要接收隐式Intent就必须要在Intent过滤器中包含"android.intent.category.DEFAULT"。

清单文件中的<intent-filter>元素使用<category>子元素列出其包含的种类，例如：

```
<intent-filter…… >
    <category android:name="android.intent.category.DEFAULT" />
    <category android:name="android.intent.category.BROWSABLE" />
    ……
</intent-filter>
```

对于一个要通过种类检测的Intent，Intent对象中的每个种类必须与过滤器中的一个相匹配，即过滤器中还能列出额外的种类。但是Intent对象中的种类都必须能够在过滤器中找到，只要有一个种类没有在过滤器列表中找到，种类检测就会失败。

> **提示**："android.intent.action.MAIN"和"android.intent.category.LAUNCHER"设置分别是标记活动开始新的任务和带到启动列表界面。它们可以包含"android.intent.category.DEFAULT"到种类列表中，也可以不包含。

（3）检测Data。

清单文件中的<intent-filter>元素使用<data>子元素列出数据，例如：

```
<intent-filter…… >
    <data android:mimeType="video/mpeg" android:scheme="http" …… />
    <data android:mimeType="audio/mpeg" android:scheme="http" …… />
    ……
</intent-filter>
```

每个<data>元素指定一个Uri和数据类型（MIME类型）。一个Uri由scheme、host、port、path几部分组成，形式如：scheme://host:port/path。

例如，下面的一行代码代表一个Uri。

```
content://com.example.project:200/folder/subfolder/etc
```

其中，scheme是"content"，host是"com.example.project"，port是200，path是"folder/subfolder/etc"。host和port一起构成Uri的凭据（authority），如果host没有指定，port也被忽略。这4个属性都是可选的，但它们之间并不都是完全独立的。要让authority有意义，scheme也必须要指定；要让path有意义，scheme和authority则都必须要指定。

<data>元素的type属性指定数据的MIME类型。Intent对象和过滤器都可以用"*"通配符匹配子类型字段，例如，"text/*"和"audio/*"可表示任何子类型。

数据检测既要检测Uri，也要检测数据类型，检测规则如下所述。

- 一个Intent对象既不包含Uri，也不包含数据类型。仅当过滤器不指定任何Uri和数据类型时，才不能通过检测，否则都能通过。
- 一个Intent对象包含Uri，但不包含数据类型。仅当过滤器不指定数据类型，同时它们的Uri匹配，才能通过检测。例如，mailto:和tel:都不指定实际数据。
- 一个Intent对象包含数据类型，但不包含Uri。仅当过滤器只包含数据类型且与Intent相同时，才能通过检测。
- 一个Intent对象既包含Uri，也包含数据类型（或数据类型能够从Uri推断出）；数据类型部分，只有与过滤器中之一匹配才算通过；Uri部分，它的Uri或出现在过滤器中，或者它有content:或file:Uri，或者过滤器没有指定Uri。

7.1.3 Intent的解析

理解Intent的关键是弄清楚Intent的两种基本用法。一种是显式的Intent，即在构造Intent对象时就指定接收者；另一种是隐式的Intent，即Intent的发送者在构造Intent对象时，并不知道也不关心接收者是谁，这有利于降低发送者和接收者之间的耦合。

对于显式的Intent，Android不需要去做解析，因为目标组件已经很明确。Android需要解析的是那些隐式的Intent，通过解析，将Intent映射给可以处理此Intent的Activity、Service或Broadcast Receiver。

Intent的解析机制主要是通过查找已注册在AndroidManifest.xml中的所有<intent-filter>定义的Intent，最终找到匹配的Component。在这个解析过程中，Android是通过Intent的action、type、category这3个属性进行判断的。判断方法如下。

- 如果Intent指明定了action，则目标组件的<intent-filter>的action列表中就必须包含有这个action，否则不能匹配。
- 如果Intent没有提供type，系统将从data中得到数据类型。与action一样，目标组件的数据类型列表中必须包含Intent的数据类型，否则不能匹配。
- 如果Intent中的数据不是content:类型的Uri，而且Intent也没有明确指定它的type，将根据Intent中数据的scheme（如http:或者mailto:）进行匹配。同理，Intent的scheme必须出现在目标组件的scheme列表中。

- 如果Intent指定了一个或多个category，这些类别必须全部出现在组建的类别列表中。例如，Intent中包含了两个类别：LAUNCHER_CATEGORY 和 ALTERNATIVE_CATEGORY，解析得到的目标组件则必须包含这两个类别。

7.1.4 Intent的实现

通过Intent可以启动另一个Activity、启动Service、发起广播Broadcast等，并可以通过它传递数据。实现方式有两种：一种是显式调用，一种是隐式调用。

1. 显式调用

Intent最常用的功能就是连接应用程序当中的各个Activity。显式Intent直接用组件的名称定义目标组件，这种方式很直接。启动一个特定Activity的核心代码如下。

Intent intent = new Intent(源Activity名.this, 目标Activity名.class);
startActivity(intent);

或者先创建ComponentName类的对象，然后将该对象设置成Intent对象的Component属性，这样应用程序即可根据该Intent的"意图"去启动指定组件。核心代码如下。

ComponentName comp = new ComponentName(源Activity名.this, 目标Activity名.class);
Intent intent = new Intent();
intent.setComponent(comp);
startActivity(intent);

或者启动不同工程项目的Activity，前提条件是被调用的类已经安装在运行的模拟器或手机上，如果知道其包名和类名，可以调用如下代码。

Intent intent = new Intent();
intent.setClassName("com.example.test","com.example.test.OtherActivity");
startActivity(intent);

开发人员往往不清楚其他应用程序的组件名称，因此，显式Intent更多是用在应用程序内部以传递消息。

【案例7-1】：在"SimpleIntentDemo"项目中通过监听命令按钮的点击动作跳到另外一屏，操作步骤如下。

步骤 01 创建项目"SimpleIntentDemo"，并创建一个Activity，命名为"SimpleIntentDemoActivity"。

步骤 02 在"res\layout"文件夹中创建名为"second.xml"的布局文件，其中只添加一个TextView，其提示信息为"这是第二屏"。

步骤 03 在src文件夹中创建名为"SecondActivity"的Activity，并重载onCreate方法，通过setContentView(R.layout.second)方法设置其布局文件。

步骤 04 在"AndroidManifest.xml"中为"SecondActivity"注册，代码如下。

```
<activity android:name=".SecondActivity"
android:label="@string/app_name">
</activity>
```

步骤 05 打开"res\layout\main.xml"文件,对它进行修改,添加一个提示信息为"跳转到第二屏"的命令按钮,并设置其id,代码为 android:id="@+id/button1" 。

编写"SimpleIntentDemoActivity",代码如下。

```
import android.app.Activity;
import android.content.Intent;
import android.os.Bundle;
import android.view.View;
import android.view.View.OnClickListener;
import android.widget.Button;

public class SimpleIntentDemoActivity extends Activity {
    public void onCreate(Bundle savedInstanceState) {
        super.onCreate(savedInstanceState);
        setContentView(R.layout.main);
        Button bt= (Button)findViewById(R.id.button1);
        bt.setOnClickListener(new ClickLis());
    }
    class ClickLis implements OnClickListener{
        public void onClick(View v) {
            Intent oneIntent=new Intent();
            oneIntent.setClass(SimpleIntentDemoActivity.this, SecondActivity.class);
            startActivity(oneIntent);
        }
    }
}
```

2.隐式调用

Intent机制更重要的在于其隐式的Intent,即Intent的发送者不指定接收者,很可能不知道也不关心接收者是谁,而是由Android框架去寻找最匹配的接收者。

(1)最简单的隐式Intent。

最简单的隐式调用就是不指定接收者,即初始化Intent对象时,只是传入参数,用Intent调用系统中的组件。例如,设定Action为Intent.ACTION_DIAL,核心代码如下。

```
Intent intent = new Intent(Intent.ACTION_DIAL);
startActivity(intent);
```

此时就会启动Android自带的实现打电话功能的Dialer程序。这里使用的构造函数的原型如下。

```
Intent(String action);
```

其中，action为Intent的常量，如Intent.ACTION_DIAL、Intent.ACTION_SEND、Intent.ACTION_VIEW等。Intent的发送者只是指定了Action，如果用户启动的Activity的描述信息正好与第三方Activity的描述信息相匹配，这个第三方的Activity就会被启动。

【案例7-2】：在"SimpleImplicitIntent"项目中通过点击监听命令按钮动作实现浏览指定的网页，操作步骤如下。

步骤 01 新建项目"SimpleImplicitIntent"，并创建一个Activity，命名为"SimpleImplicitIntentActivity"。

步骤 02 打开"res\layout"文件夹，修改其中的"main.xml"布局文件，其中只添加一个Button，其提示信息为"打开网页"，并设置其id，代码为 android:id="@+id/button" 。

步骤 03 打开"SimpleImplicitIntentActivity.java"文件，修改为如下代码。

```java
package com.chapt6;
import android.app.Activity;
import android.content.Intent;
import android.net.Uri;
import android.os.Bundle;
import android.view.View;
import android.view.View.OnClickListener;
import android.widget.Button;

public class SimpleImplicitIntentActivity extends Activity {
    public void onCreate(Bundle savedInstanceState) {
        super.onCreate(savedInstanceState);
        setContentView(R.layout.main);
        Button bt = (Button) findViewById(R.id.button);
        bt.setOnClickListener(new OnClickListener() {
            public void onClick(View v) {
                Uri myuri = Uri.parse("http://www.baidu.com");
                Intent intent = new Intent(Intent.ACTION_VIEW, myuri);
                startActivity(intent);
            }
```

```
        });
    }
}
```

在上述代码中可以修改Intent相关的语句，例如，使其可通过Intent播放音频文件，相关的修改语句如下。

```
Uri myuri=Uri.parse("file:///sdcard/song.mp3");
Intent intent = new Intent(Intent.ACTION_VIEW);
intent.setDataAndType(myuri, "audio/mp3");
startActivity(intent);
```

（2）增加接收者的隐式Intent。

接收者如果希望能够接收某些Intent，需要在"AndroidManifest.xml"文件中增加Activity的声明，并设置对应的Intent过滤器和Action，才能被Android的应用程序框架匹配。

【案例7-3】：通过点击监听命令按钮的动作，实现在"Phone"（系统自带的打电话程序）和"打开自定的拨号界面"之间的选择，用户选择不同的按钮会跳到不同的界面，效果如图7-1所示。

图 7-1　具有接收者的隐式 Intent

操作步骤如下。

步骤01 新建项目"DialDemo"，并创建一个Activity，命名为"MyDialActivity"。

步骤02 打开"res\layout"文件夹，修改其中的"main.xml"布局文件，在其中添加一个Button，其提示信息为"dialtel"，并设置其id，代码为 `android:id="@+id/button"` 。

步骤03 修改"MyDialActivity.java"文件，并重载onCreate方法，代码如下。

```
package com.chapt6;
import android.app.Activity;
import android.content.Intent;
import android.net.Uri;
import android.os.Bundle;
import android.view.View;
import android.view.View.OnClickListener;
```

```java
import android.widget.Button;

public class MyDialActivity extends Activity {
    public void onCreate(Bundle savedInstanceState) {
        super.onCreate(savedInstanceState);
        setContentView(R.layout.main);
        Button bt = (Button) findViewById(R.id.button);
        bt.setOnClickListener(new OnClickListener() {
            public void onClick(View v) {
                Intent intent = new Intent(Intent.ACTION_DIAL);
                startActivity(intent);
            }
        });
    }
}
```

步骤 04 在src文件夹中新建一个名为"DialTelActivity"的Activity，并存放在com.chapt6包中。

步骤 05 打开"res\values\string.xml"文件，修改代码。

```xml
<resources>
    <string name="hello">Hello World, MyDialActivity!</string>
    <string name="app_name">DialDemo</string>
    <string name="dial">dialtel</string>
    <string name="title_activity_my_dial">打开自定的拨号界面</string>
</resources>
```

步骤 06 修改"AndroidManifest.xml"文件，将"DialTelActivity"的声明部分改为：

```xml
<activity android:name="com.chapt6.DialTelActivity"
    android:label="@string/title_activity_my_dial">
    <intent-filter>
        <action android:name="android.intent.action.DIAL" />
        <category android:name="android.intent.category.DEFAULT" />
    </intent-filter>
</activity>
```

针对Intent.ACTION_DIAL，Android框架找到了两个符合条件的Activity，因此，它将这两个Activity分别列出供用户选择。当选择"Phone"时，打开系统拨打电话的界面；当选择"打开自定的拨号界面"时，就会打开用户自定义的拨打电话界面，这里没有设置其布局，仅仅为

空界面。

在"AndroidManifest.xml"文件中增加的两行代码是:

```
<action android:name="android.intent.action.DIAL" />
<category android:name="android.intent.category.DEFAULT" />
```

这两行修改了原来的Intent过滤器,这样Activity就能够接收到发送的Intent。Intent发送者设定Action来说明将要进行的动作,而Intent的接收者在"AndroidManifest.xml"文件中通过设定Intent过滤器来声明自己能接收哪些Intent。

7.1.5 Intent中传递数据

Intent除了具有定位目标组件的功能外,还有一个功能就是传递数据信息。Intent之间传递数据有两种常用的方法:一种是通过data属性,另一种是通过extra属性。data属性是一种Uri,它可以指向HTTP、FTP等网络地址,也可以指向ContentProvider提供的资源。通过调用Intent的setData方法放入数据,使用getData方法取出数据。

如果需要启动Android内置的浏览器,可使用下面的代码将网址通过data属性传递给它。

```
Intent intent = new Intent(Intent.ACTION_VIEW);
intent.setData(Uri.parse("http://www.baidu.com"));
startActivity(intent);
```

如果需要传递数据对象,则需要使用extra属性。Intent提供了多个重载的方法来"携带"额外的数据。

- **putExtra(String name, Xxx value)**:向Intent中放入Xxx类型的数据。
- **putIntegerArrayListExtra(String name, ArrayList<Integer> value)**:向Intent中放入ArrayList类型的数据。
- **putStringArrayListExtra(String name, ArrayList<String> value)**:向Intent中放入ArrayList类型的数据。
- **putExtras(Bundle extras)**:在Intent中通过Bundle对象传递数据。

如何获取传递的数据呢?先使用getIntent方法得到上个Activity传递过来的Intent内容,然后根据数据的类型使用Intent的getXxxExtra(String key)方法获取相应的数据。

利用Bundle是一种比较方便的方法。Android中的Bundle是一种类似于哈希表的数据结构,是一种名值对。可以将各种基本类型的数据保存在Bundle类中打包传输。在Bundle中定义了putXxx(String key, Xxx value)方法,向Bundle中放入int、long等各种类型的数据。

为了取出Intent中携带的数据,Intent中提供了如下方法。

- **getExtras方法**:获取一个Bundle对象,然后使用Bundle的get方法获取数据的值。
- **getXxx(String key)方法**:从Bundle中取出Xxx类型的数据。
- **getXxx(String key, Xxx defaultValue)方法**:从Bundle中取出Xxx类型的数据,如果取不

到，则使用defaultValue。

【案例7-4】：创建用户登录界面，让用户输入用户名和密码，点击"登录"命令按钮时，跳转到另一个界面，显示欢迎信息，并把登录信息显示到当前界面上，效果如图7-2所示。

图7-2 运行效果

操作步骤如下。

步骤 01 新建项目"IntentBundleDemo"，并创建一个Activity，命名为"IntentBundleActivity"。

步骤 02 打开"res\values\strings.xml"，修改为如下代码。

```
<?xml version="1.0" encoding="utf-8"?>
<resources>
    <string name="app_name">IntentBundleDemo</string>
    <string name="username">用户名</string>
    <string name="password">密码</string>
    <string name="login">登录</string>
    <string name="cancel">取消</string>
</resources>
```

步骤 03 打开"res\layout"文件夹，修改其中的"main.xml"布局文件，代码如下。

```
<?xml version="1.0" encoding="utf-8"?>
<LinearLayout xmlns:android="http://schemas.android.com/apk/res/android"
    android:orientation="vertical"
    android:layout_width="fill_parent"
    android:layout_height="fill_parent">
    <LinearLayout
        android:layout_width="fill_parent"
        android:layout_height="wrap_content"
        android:gravity="center">
        <TextView
            android:layout_width="0dip"
```

```xml
        android:layout_height="wrap_content"
        android:layout_weight="1"
        android:text="@string/username" />
    <EditText
        android:id="@+id/username"
        android:layout_width="0dip"
        android:layout_height="wrap_content"
        android:layout_weight="1" />
</LinearLayout>
<LinearLayout
    android:layout_width="fill_parent"
    android:layout_height="wrap_content"
    android:gravity="center">
    <TextView
        android:layout_width="0dip"
        android:layout_height="wrap_content"
        android:layout_weight="1"
        android:text="@string/password" />
    <EditText
        android:id="@+id/userpassword"
        android:layout_width="0dip"
        android:layout_height="wrap_content"
        android:layout_weight="1" />
</LinearLayout>
<LinearLayout
    android:layout_width="fill_parent"
    android:layout_height="wrap_content"
    android:gravity="center">
    <Button
        android:layout_width="wrap_content"
        android:layout_height="wrap_content"
        android:id="@+id/buttonlogin"
        android:text="@string/login" />
    <Button
        android:layout_width="wrap_content"
        android:layout_height="wrap_content"
        android:id="@+id/buttoncancel"
```

```
            android:text="@string/cancel" />
    </LinearLayout>
</LinearLayout>
```

步骤 04 修改"IntentBundleActivity.java"文件，并重载onCreate方法，代码如下。

```java
package com.chapt6;
import android.app.Activity;
import android.content.Intent;
import android.os.Bundle;
import android.view.View;
import android.view.View.OnClickListener;
import android.widget.Button;
import android.widget.TextView;

public class IntentBundleActivity extends Activity {
    public void onCreate(Bundle savedInstanceState) {
        super.onCreate(savedInstanceState);
        setContentView(R.layout.main);
        Button bt1 = (Button) findViewById(R.id.buttonlogin);
        bt1.setOnClickListener(new OnClickListener() {

            public void onClick(View v) {
                Bundle data = new Bundle();
                // 向Bundle中绑定数据，以名值对的形式
                TextView username = (TextView) findViewById(R.id.username);
                TextView userpassword = (TextView) findViewById(R.id.userpassword);
                data.putString("name", username.getText().toString());
                data.putString("password", userpassword.getText().toString());
                Intent intent = new Intent(IntentBundleActivity.this, ShowResultActivity.class);
                // 把Bundle绑定到Intent中
                intent.putExtras(data);
                startActivity(intent);
            }
        });
    }
}
```

步骤 05 打开"res\layout"文件夹中新建的布局文件"showresultlayout.xml",代码如下。

```xml
<?xml version="1.0" encoding="utf-8"?>
<LinearLayout
    xmlns:android="http://schemas.android.com/apk/res/android"
    android:layout_width="match_parent"
    android:layout_height="match_parent">
    <TextView
        android:layout_width="match_parent"
        android:layout_height="match_parent"
        android:id="@+id/showresult" >
    </TextView>
</LinearLayout>
```

步骤 06 在src文件夹中定义一个名为"ShowResultActivity"的Activity文件,并重载onCreate方法,代码如下。

```java
package com.chapt6;
import android.app.Activity;
import android.content.Intent;
import android.os.Bundle;
import android.view.View;
import android.view.View.OnClickListener;
import android.widget.Button;
import android.widget.TextView;

public class ShowResultActivity extends Activity {
    protected void onCreate(Bundle savedInstanceState) {
        super.onCreate(savedInstanceState);
        setContentView(R.layout.showresultlayout);
        TextView show=(TextView)findViewById(R.id.showresult);
        Intent intent=getIntent();
        //获取Intent中绑定的Bundle
        Bundle result=intent.getExtras();
        String username=result.getString("name");
        String userpassword=result.getString("password");
        show.setText("欢迎使用！您的用户名为："+username+"您的密码为："+userpassword);
    }
}
```

步骤 07 在"AndroidManifest.xml"文件中为ShowResultActivity进行注册,添加如下代码。

```xml
<activity
    android:name=".ShowResultActivity"
    android:label="@string/app_name" >
</activity>
```

7.1.6 在Intent中传递复杂对象

Android的Intent之间传递对象有两种方法:一种是Bundle.putSerializable(Key,Object),另一种是Bundle.putParcelable(Key,Object)。方法中的Object要满足一定的条件,前者实现了Serializable接口,而后者实现了Parcelable接口。

Android设计团队认为Java中的序列化太慢,难以满足Android进程间通信的需求,所以他们构建了Parcelable解决方案。Parcelable接口要求显式地序列化类的成员,但最终序列化对象的速度将快很多。在Android运行环境中推荐使用Parcelable接口,它不但可以利用Intent传递,还可以在远程方法调用中使用。

实现Parcelable接口需要先满足3个方法。

(1) writeToParcel方法。

该方法将类的数据写入外部提供的Parcel中,其实现格式为:

writeToParcel (Parcel dest, int flags)

(2) describeContents方法。

返回内容描述信息的资源ID,默认返回0。

(3) 静态的Parcelable.Creator<T>接口。

本接口有两个方法:

- **createFromParcel(Parcel in)**:实现从in中创建出类的实例的功能。
- **newArray(int size)**:创建一个类型为T,长度为size的数组。

【案例7-5】:使用Serializable和Parcelable接口传递对象,操作步骤如下。

步骤 01 创建项目"IntentObjectDemo",并包含一个名为"IntentObjectActivity"的Activity。

步骤 02 在src文件夹中创建类"SerializableUser",并实现Serializable接口,代码如下。

```java
package com.chapt6;
import java.io.Serializable;
public class SerializableUser implements Serializable {
    private String userName;
    private String passWord;
    public String getUserName() {
        return userName;
```

```
    }
    public void setUserName(String userName) {
        this.userName = userName;
    }
    public String getPassWord() {
        return passWord;
    }
    public void setPassWord(String passWord) {
        this.passWord = passWord;
    }
    public SerializableUser(String userName, String passWord) {
        this.userName = userName;
        this.passWord = passWord;
    }
}
```

步骤 03 在src文件夹中创建类"ParcelableUser",并实现Parcelable接口,代码如下。

```
package com.chapt6;
import android.os.Parcel;
import android.os.Parcelable;

public class ParcelableUser implements Parcelable {
    private String userName;
    private String password;
    public ParcelableUser() {
    }
    public ParcelableUser(String userName, String password) {
        this.userName = userName;
        this.password = password;
    }
    public String getUserName() {
        return userName;
    }
    public void setUserName(String userName) {
        this.userName = userName;
    }
    public String getPassword() {
```

```java
        return password;
    }
    public void setPassword(String password) {
        this.password = password;
    }
    public int describeContents() {
        return 0;
    }
    public void writeToParcel(Parcel p, int arg1) {
        p.writeString(userName);
        p.writeString(password);
    }

    public static final Parcelable.Creator<ParcelableUser> CREATOR=new Creator<ParcelableUser>(){
        public ParcelableUser createFromParcel(Parcel source) {
            ParcelableUser parcelableUser = new ParcelableUser();
            parcelableUser.userName = source.readString();
            parcelableUser.password = source.readString();
            return parcelableUser;
        }
        public ParcelableUser[] newArray(int size) {
            return new ParcelableUser[size];
        }
    }
}
```

步骤 04 打开 "res\layout\main.xml"，代码修改如下。

```xml
<?xml version="1.0" encoding="utf-8"?>
<LinearLayout xmlns:android="http://schemas.android.com/apk/res/android"
    android:orientation="vertical"
    android:layout_width="fill_parent"
    android:layout_height="fill_parent"
    >
<Button
    android:id="@+id/button1"
    android:layout_width="wrap_content"
    android:layout_height="wrap_content"
```

```xml
            android:text="传递Serializable对象"
            android:onClick="sendData" />
    <Button
            android:id="@+id/button2"
            android:layout_width="wrap_content"
            android:layout_height="wrap_content"
            android:text="传递Parcelable对象"
            android:onClick="sendData" />
</LinearLayout>
```

步骤 05 修改"IntentObjectActivity.java"文件，并重载onCreate方法，代码如下。

```java
package com.chapt6;
import android.app.Activity;
import android.content.Intent;
import android.os.Bundle;
import android.view.View;

public class IntentObjectActivity extends Activity {
    public void onCreate(Bundle savedInstanceState) {
        super.onCreate(savedInstanceState);
        setContentView(R.layout.main);
    }
    public void sendData(View view){
        switch(view.getId()){
        case R.id.button1:
            SerializableUser sUser = new SerializableUser("Admin", "123456");
            Intent intent = new Intent(this,ReceiveObjectActivity.class);
            Bundle bundle = new Bundle();
            bundle.putInt("type", 1);
            bundle.putSerializable("serial", sUser);
            intent.putExtras(bundle);
            startActivity(intent);
            break;
        case R.id.button2:
            ParcelableUser pUser = new ParcelableUser("User", "123456");
            Intent intent1 = new Intent(this,ReceiveObjectActivity.class);
            Bundle bundle1 = new Bundle();
```

```
            bundle1.putInt("type", 2);
            bundle1.putParcelable("parcel", pUser);
            intent1.putExtras(bundle1);
            startActivity(intent1);
            break;
        }
    }
}
```

步骤 06 打开"res\layout"文件夹，新建布局文件"objectreceiver.xml"，代码如下。

```
<?xml version="1.0" encoding="utf-8"?>
<LinearLayout
    xmlns:android="http://schemas.android.com/apk/res/android"
    android:layout_width="match_parent"
    android:layout_height="match_parent">
    <TextView
        android:layout_width="match_parent"
        android:layout_height="match_parent"
        android:id="@+id/showresult" >
    </TextView>
</LinearLayout>
```

步骤 07 在src文件夹中定义一个名为"ReceiveObjectActivity"的Activity，并重载onCreate方法，代码如下。

```
package com.chapt6;
import android.app.Activity;
import android.os.Bundle;
import android.widget.TextView;

public class ReceiveObjectActivity extends Activity {
    protected void onCreate(Bundle savedInstanceState) {
        super.onCreate(savedInstanceState);
        setContentView(R.layout.objectreceiver);
        TextView tv = (TextView) findViewById(R.id.showresult);
        Bundle bundle = getIntent().getExtras();
        int type = bundle.getInt("type");
        if (type == 1) {
```

```
            SerializableUser serializableUser = (SerializableUser) getIntent() .getSerializableExtra("serial");
            tv.setText(serializableUser.getUserName() + "\n" + serializableUser.getPassWord());
        } else {
            ParcelableUser parcelableUser = (ParcelableUser) getIntent() .getParcelableExtra("parcel");
            tv.setText(parcelableUser.getUserName() + "\n"+ parcelableUser.getPassword());
        }
    }
}
```

步骤 08 修改"AndroidManifest.xml"文件，为ReceiveObjectActivity类注册。

```
<activity
    android:name=".ReceiveObjectActivity"
    android:label="@string/app_name">
</activity>
```

7.2　ContentProvider

系统的多个应用程序之间，通常需要共享与交换数据，但Intent只适用于传递数据量小的场合，不适用于大数据文件之间的共享与交换。为此，Android还提供了ContentProvider类，以便实现数据共享。

7.2.1　ContentProvider简介

ContentProvider是一个抽象类，是一种特殊的存储数据类型，它提供了一套标准的接口来获取和操作数据。Android自身也提供了现成的ContentProvider:Contents组件，可以把数据封装到ContentProvider中，从而使这些数据可以供其他应用程序共享，这样就搭建起了所有应用程序之间数据交换的桥梁。

当应用继承ContentProvider类，并重载该类用于提供数据和存储数据的方法，就可以向其他应用共享其数据。虽然使用其他方法也可以对外共享数据，但数据访问方式会因数据存储方式的不同而不同，例如，采用文件方式对外共享数据，需要使用文件操作读写数据；采用SharedPreferences共享数据，需要使用SharedPreferences API读写数据。使用ContentProvider共享数据的好处是统一了数据的访问方式。

ContentProvider类实现了一组标准的方法接口，从而能够让其他的应用程序保存或读取此ContentProvider的各种数据类型。在程序内可以通过实现ContentProvider的抽象接口将数据显示出来，外界通过统一的接口实现数据的增、删、改、查。

当应用需要通过ContentProvider对外共享数据时，首先需要继承ContentProvider并重载以下方法。

```
public class UserProvider extends ContentProvider{
    public boolean onCreate()
    public Uri insert(Uri uri, ContentValues values)
    public int delete(Uri uri, String selection, String[] selectionArgs)
    public int update(Uri uri, ContentValues values, String selection, String[] selectionArgs)
    public Cursor query(Uri uri, String[] projection, String selection, String[] selectionArgs, String sortOrder)
    public String getType(Uri uri)
}
```

接下来需要在"AndroidManifest.xml"文件中使用<provider>对该ContentProvider进行配置,为了方便其他应用找到该ContentProvider,ContentProvider采用了authorities(主机名/域名)对它进行唯一标识,即可以把ContentProvider看作是一个网站,authorities就是它的域名。

```
<provider android:name=".UserProvider"
    android:authorities="com.chapter.userprovider"/>
```

7.2.2 Uri、UriMatcher、ContentUris和ContentResolver类简介

使用ContentProvider,Uri起着关键作用,因为它决定了访问哪个ContentProvider。

1. Uri

Uri代表了要操作的数据,它主要包含两部分信息:一是需要操作的ContentProvider,二是对ContentProvider中的什么数据进行操作。Uri由3部分组成,如图7-3所示。

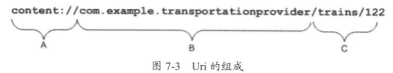

图7-3 Uri 的组成

图中A部分表示ContentProvider的scheme,它已由Android所规定,内容为:content://。

图中B部分表示主机名(也称authorities),是一个ContentProvider的唯一标识,外部调用者可以通过该标识找到这个ContentProvider。

图中C部分表示路径(也称path),表示要操作的数据的所在位置。构建路径时应该根据业务而定。

例如:

(1)要访问"userinfo"表中id为2的记录,构建的路径为:/userinfo/2。

(2)要操纵"userinfo"表中id为2的记录的"username"字段,构建的路径为:userinfo/2/username。

(3)要访问"userinfo"表中的所有记录,构建的路径为:/userinfo。

Uri类中的parse方法，可以把一个字符串转换成Uri，代码如下。

Uri uri = Uri.parse("content://com.chapt6.userprovider/userinfo");

在使用ContentProvider时，几乎都会用到Uri。如果是自定义的ContentProvider，通常将Uri定义为常量，从而在简化开发的同时也提高了程序的可维护性。

Uri代表了要操作的数据，所以需要解析Uri以获取数据。Android系统提供了UriMatcher类和ContentUris类来操作Uri。

2. UriMatcher

该类用于匹配Uri，它的用法如下。

（1）注册需要匹配Uri的路径。

```
// UriMatcher.NO_MATCH表示不匹配任何路径的返回码。
UriMatcher oneMatcher = new UriMatcher(UriMatcher.NO_MATCH);

//如果match方法匹配content:// com.chapt6.userprovider/userinfo路径，返回匹配码1。添加需要匹配的
//uri，如果匹配就会返回匹配码。
oneMatcher.addURI("com.chapt6.userprovider ", " userinfo", 1);
//如果match方法匹配 content:// com.chapt6.userprovider/userinfo/#路径，返回匹配码为2，其中#号
//表示通配符。
oneMatcher.addURI("com.chapt6.userprovider ", " userinfo /#", 2);
```

（2）Uri注册完后，通过uriMatcher.match(uri)方法对Uri匹配，若匹配成功返回对应的匹配码。例如：oneMatcher.match(Uri.parse("content:// com.chapt6.userprovider/userinfo/2"))，返回的匹配码为2。

3. ContentUris

该类用于获取Uri路径后面的id部分。两个常用的方法是withAppendedId(uri, id)和parseId(uri)，分别是为路径加上id和获取id。例如：

```
Uri uri =Uri.parse("content:// com.chapt6.userprovider/userinfo/");
Uri resultUri =ContentUris.withAppendedId(uri, 5);
//该语句执行后生成content:// com.chapt6.userprovider/userinfo/5的Uri
Uri uri =Uri.parse("content:// com.provider.userprovider/userinfo/5");
Long userid = ContentUris.parseId(uri);//获取的id为5
```

4. ContentResolver

ContentResolver是通过ContentProvider获取与应用程序共享的数据，当外部应用需要访问ContentProvider中的数据时，可以使用ContentResolver类来完成对数据的增、删、改及查询操

作。使用时，首先通过getContext().getContentResolver()方法获取ContentResolver对象，然后再通过ContentResolver对象的insert、delete、update、query方法操作数据。

7.2.3 自定义ContentProvider

本节将通过一个示例来展示如何自定义ContentProvider。

【案例7-6】：通过ContentProvider读取和获取数据库的信息，并把结果通过Logcat显示输出。操作步骤如下。

步骤 01 创建项目"DefineContentProvider"，并包含一个名为"MainActivity"的Activity。

步骤 02 在src文件夹中创建类DataBaseHelper继承自SQLiteOpenHelper。在该类中创建的数据库名为"userdb.db"，数据库中的表名为"userinfo"。代码如下。

```
package com.chapt6;
import android.content.Context;
import android.database.sqlite.SQLiteDatabase;
import android.database.sqlite.SQLiteDatabase.CursorFactory;
import android.database.sqlite.SQLiteOpenHelper;
import android.util.Log;

public class DataBaseHelper extends SQLiteOpenHelper {
    public static final String DB_NAME = "userdb.db";
    public static final String TABLENAME = "userinfo";
    public static final int DB_VERSION = 1;
    public static final String CREATETABLE = "create table " + TABLENAME
        + "(_id integer primary key,username text, userpassword text);";

    public DataBaseHelper(Context context) {
        super(context, DB_NAME, null, DB_VERSION);
    }
    public void onCreate(SQLiteDatabase db) {
        db.execSQL(CREATETABLE);
    }

    public void onUpgrade(SQLiteDatabase db, int oldVersion, int newVersion) {
        Log.i("Database update......", "Update database from " + oldVersion+ " to " + newVersion);
        // 删除旧表
        db.execSQL("drop table if it exists " + TABLENAME);
        // 创建新表
```

```
        onCreate(db);
    }
}
```

步骤 03 在src文件夹中创建类UserProvider继承自ContentProvider，在该类中需要实现6个方法，代码如下。

```java
package com.chapt6;
import android.content.ContentProvider;
import android.content.ContentUris;
import android.content.ContentValues;
import android.content.UriMatcher;
import android.database.Cursor;
import android.database.sqlite.SQLiteDatabase;
import android.net.Uri;

public class UserProvider extends ContentProvider {
    private DataBaseHelper dbh = null;
    // 1.发布Content Provider的Uri地址
    private static final String AUTHORITY = "com.chapt6.userprovider";
    public static final Uri CONTENT_URI = Uri
            .parse("content://com.chapt6.userprovider/userinfo");

    // 2.注册需要匹配的Uri
    private static UriMatcher uriMatcher = new UriMatcher(UriMatcher.NO_MATCH);
    static {
        uriMatcher.addURI(AUTHORITY, "userinfo", 1);
        uriMatcher.addURI(AUTHORITY, "userinfo/#", 2);
    }

    // 该方法在ContentProvider创建后就会被调用，在其他应用第1次访问它时才会被创建
    public boolean onCreate() {
    // 3.实例化dbh
        dbh = new DataBaseHelper(getContext());
        return false;
    }
    // 4.返回当前Uri所代表的MIME类型数据
    // 该方法用于返回当前Uri所代表的MIME类型数据
```

```java
public String getType(Uri uri) {
    switch (uriMatcher.match(uri)) {
    case 1:
        return "vnd.android.cursor.dir/userinfo";
    case 2:
        return "vnd.android.cursor.item/userinfo";
    }
    return null;
}
// 5.实现删除方法
// 该方法用于供外部应用从ContentProvider中删除数据
public int delete(Uri uri, String selection, String[] selectionArgs) {
    SQLiteDatabase db = dbh.getWritableDatabase();
    // 已经删除的记录数量
    int num = 0;
    switch (uriMatcher.match(uri)) {
    case 1:
        num = db.delete("userinfo", selection, selectionArgs);
        break;
    case 2:
        // 获取ID
        long id = ContentUris.parseId(uri);
        // 在selection上增加条件_id=id
        if (selection == null) {
            selection = "_id=" + id;
        } else {
            selection = "_id=" + id + " and (" + selection + ")";
        }
        num = db.delete("userinfo", selection, selectionArgs);
        break;
    default:
        break;
    }
    // 通知所有的观察者，数据集已经改变
    getContext().getContentResolver().notifyChange(uri, null);
    return num;
}
```

```java
// 6.实现插入方法
// 该方法用于供外部应用往ContentProvider中添加数据
public Uri insert(Uri uri, ContentValues values) {
    SQLiteDatabase db = dbh.getWritableDatabase();
    long id = db.insert("userinfo", null, values);
    if (id > -1) {// 插入数据成功
        // 构建新插入行的Uri
        Uri insertUri = ContentUris.withAppendedId(CONTENT_URI, id);
        // 通知所有的观察者，数据集已经改变
        getContext().getContentResolver().notifyChange(insertUri, null);
        return insertUri;
    }
    return null;
}
// 7.实现查询方法
// 该方法用于供外部应用从ContentProvider中获取数据
public Cursor query(Uri uri, String[] projection, String selection,
        String[] selectionArgs, String sortOrder) {
    SQLiteDatabase db = dbh.getReadableDatabase();
    Cursor cursor = null;
    switch (uriMatcher.match(uri)) {
    // 查询所有行
        case 1:
            cursor = db.query("userinfo",       // 表名
                    null,                        // 列的数组，null代表所有列
                    selection,                   // where条件
                    selectionArgs,               // where条件的参数值的数组
                    null,                        // 分组
                    null,                        // having
                    sortOrder);                  // 排序规则
            break;
        // 查询指定id的行
        case 2:
            // 获取id
            long id = ContentUris.parseId(uri);
            // 在selection上增加条件 _id=id
            if (selection == null) {
```

```java
            selection = "_id=" + id;
        } else {
            selection = "_id=" + id + " and (" + selection + ")";
        }
        cursor = db.query("userinfo",           // 表名
            null,                                // 列的数组，null代表所有列
            selection,                           // where条件
            selectionArgs,                       // where条件的参数值的数组
            null,                                // 分组
            null,                                // having
            sortOrder);                          // 排序规则
        break;
    default:
        break;
    }
    return cursor;
}
// 8.实现修改方法
// 该方法用于外部应用更新ContentProvider中的数据
public int update(Uri uri, ContentValues values, String selection,
        String[] selectionArgs) {
    SQLiteDatabase db = dbh.getWritableDatabase();
    // 已经修改的记录数量
    int num = 0;
    switch (uriMatcher.match(uri)) {
    case 1:
        num = db.update("userinfo", values, selection, selectionArgs);
        break;
    case 2:
        // 获取id
        long id = ContentUris.parseId(uri);
        // 在selection上增加条件_id=id
        if (selection == null) {
            selection = "_id=" + id;
        } else {
            selection = "_id=" + id + " and (" + selection + ")";
        }
```

```
        num = db.update("userinfo", values, selection, selectionArgs);
        break;
    default:
        break;
    }
    // 通知所有的观察者，数据集已经改变
    getContext().getContentResolver().notifyChange(uri, null);
    return num;
    }
}
```

> **提示**：如果操作的数据属于集合类型，那么MIME类型的字符串应该以"vnd.android.cursor.dir/"开头。如果要得到userinfo所有记录的Uri为"content://com.chapt6.userprovider/userinfo"，那么返回的MIME类型字符串应该为："vnd.android.cursor.dir/userinfo"。
> 如果要操作的数据属于非集合类型数据，那么MIME类型字符串应该以"vnd.android.cursor.item/"开头。如果要得到id为2的userinfo记录，Uri为"content://com.chapt6.userprovider/userinfo/2"，那么返回的MIME类型字符串为："vnd.android.cursor.item/userinfo"。

步骤 04 注册UserProvider。在"AndroidManifest.xml"文件中使用<provider>对ContentProvider进行配置，代码如下。

```
<provider
    android:name="com.chapt6.UserProvider"
    android:authorities="com.chapt6.userprovider"
    android:exported="true">
</provider>
```

步骤 05 访问UserProvider。打开src文件夹中的"MainActivity.java"文件，在MainActivity的onCreate方法中增加代码，通过UserProvider进行数据的增、删、改、查等操作，并将结果通过Logcat显示输出。代码如下。

```
package com.chapt6;
import android.app.Activity;
import android.content.ContentResolver;
import android.content.ContentUris;
import android.content.ContentValues;
import android.database.Cursor;
import android.net.Uri;
```

```java
import android.os.Bundle;
import android.util.Log;

public class MainActivity extends Activity {
    public void onCreate(Bundle savedInstanceState) {
        super.onCreate(savedInstanceState);
        setContentView(R.layout.main);
        ContentResolver cr = getContentResolver();
        // 增加记录
        ContentValues values = new ContentValues();
        values.put("username", "admin");
        values.put("userpassword", "123456");
        cr.insert(UserProvider.CONTENT_URI, values);
        values.clear();
        values.put("username", "zhangsan");
        values.put("userpassword", "666666");
        cr.insert(UserProvider.CONTENT_URI, values);
        // 查询所有记录
        Cursor cursor = cr.query(UserProvider.CONTENT_URI, null, null, null, null);
        Log.i("after inserted", "----------------------------------------");
        while (cursor.moveToNext()) {
            Log.i("after inserted", "id:" + cursor.getString(0) + " username:"
                + cursor.getString(1) + " userpassword:"+ cursor.getString(2));
        }
        cursor.close();
        // 修改记录
        values.clear();
        values.put("username", "lisi");
        // 构建的Uri为"content://com.chapt6.userprovider/userinfo/2"
        Uri uri = ContentUris.withAppendedId(UserProvider.CONTENT_URI, 2);
        // 修改id为2的记录
        cr.update(uri, values, null, null);
        // 查询id为2的记录
        cursor = cr.query(uri, null, null, null, null);
        Log.i("after updated","--------------------------------------------");
        while (cursor.moveToNext()) {
            Log.i("after updated","id:" + cursor.getString(0) + "username:"
```

```
        + cursor.getString(1) + " userpassword:"+ cursor.getString(2));
    }
    cursor.close();
    // 删除id为2的记录
    cr.delete(uri, null, null);
    // 查询记录
    cursor = cr.query(UserProvider.CONTENT_URI, null, null, null, null);
    Log.i("after deleted","-------------------------------------------------");
    while (cursor.moveToNext()) {
        Log.i("after deleted","id:" + cursor.getString(0) + "username:"
        + cursor.getString(1) + " userpassword:"+ cursor.getString(2));
    }
    cursor.close();
  }
}
```

7.2.4 系统ContentProvider

Android提供了一些主要数据类型的ContentProvider，如音频、视频、图片和私人通讯录等，可以在android.provider包中找到一些android提供的ContentProvider，并可查询它们包含的数据，当然前提是已获得适当的读取权限。

【案例7-7】：通过ContentProvider读取短信信息，并将结果通过一个TextView来显示。

Telephony Provider提供了电话、短信和彩信相关数据的共享，可以通过它访问手机中的短信信息。该数据库保存的路径为"\data\data\com.android.providers.telephony\databases\mmssms.db"。在"mmssms.db"数据库中，短信存储在"sms"表中。sms表结构如图7-4所示。

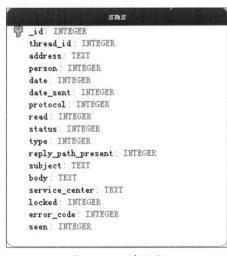

图7-4 sms表结构

其中主要字段的说明如下。
- **_id**：表示短信的id。
- **thread_id**：表示该短信所属会话的id，每个会话代表和一个联系人之间短信的群组。
- **address**：表示该短信发件人的地址，即发件人的手机号，如+861366666××××。
- **person**：表示该短信的发件人，返回的是一个数字，即联系人在列表中的序号。陌生人为null。
- **date**：表示该短信的接收日期。
- **date_sent**：表示该短信的发送日期。
- **protocol**：表示协议，0表示SMS_RPOTO，1表示MMS_PROTO。
- **read**：表示该短信是否已读。
- **type**：表示该短信的类型，例如，1表示接收类型，2表示发送类型，3表示草稿类型。
- **body**：表示短信的内容。

通过查看API和源文件"\sources\android-19\android\provider\Telephony.java"，可以发现主要有以下Uri。
- **content://sms/**：表示所有短信。
- **content://sms/inbox**：表示收件箱。
- **content://sms/sent**：表示已发送。
- **content://sms/draft**：表示草稿。
- **content://sms/outbox**：表示发件箱。
- **content://sms/failed**：表示发送失败。
- **content://sms/queued**：表示待发送列表。

操作步骤如下。

步骤 01 创建项目"ReadSmsDemo"，并包含一个名为"MainActivity"的Activity类。

步骤 02 修改"src\MainActivity.java"文件，增加短信读取代码，代码如下。

```
package com.chapt6;
import java.text.SimpleDateFormat;
import java.util.Date;
import java.util.Locale;
import android.app.Activity;
import android.database.Cursor;
import android.database.sqlite.SQLiteException;
import android.net.Uri;
import android.os.Bundle;
import android.util.Log;
import android.widget.ScrollView;
```

```java
import android.widget.TextView;

public class MainActivity extends Activity {
    public void onCreate(Bundle savedInstanceState) {
        super.onCreate(savedInstanceState);
        TextView tv = new TextView(this);
        tv.setText(getSmsInPhone());
        ScrollView sv = new ScrollView(this);
        sv.addView(tv);
        setContentView(sv);
    }
    public CharSequence getSmsInPhone() {
        final String SMS_URI_ALL = "content://sms/";
        StringBuilder smsBuilder = new StringBuilder();
        try {
            Uri uri = Uri.parse(SMS_URI_ALL);
            String[] projection = new String[] { "_id", "address", "person", "body", "date", "type" };
            Cursor cur = getContentResolver().query(uri, projection, null, null, " date desc");
            if (cur.moveToFirst()) {
                int index_Address = cur.getColumnIndex("address");
                int index_Person = cur.getColumnIndex("person");
                int index_Body = cur.getColumnIndex("body");
                int index_Date = cur.getColumnIndex("date");
                int index_Type = cur.getColumnIndex("type");
                do {
                    String strAddress = cur.getString(index_Address);
                    int intPerson = cur.getInt(index_Person);
                    String strbody = cur.getString(index_Body);
                    long longDate = cur.getLong(index_Date);
                    int intType = cur.getInt(index_Type);
                    SimpleDateFormat dateFormat = new SimpleDateFormat(
                        "yyyy-MM-dd hh:mm:ss", Locale.US);
                    Date d = new Date(longDate);
                    String strDate = dateFormat.format(d);
                    String strType = "";
                    if (intType == 1) {
                        strType = "接收";
```

```
            } else if (intType == 2) {
                strType = "发送";
            } else {
                strType = "null";
            }
            smsBuilder.append(strAddress + ", ");
            smsBuilder.append(intPerson + ", ");
            smsBuilder.append(strbody + ", ");
            smsBuilder.append(strDate + ", ");
            smsBuilder.append(strType);
            smsBuilder.append("\n");
        } while (cur.moveToNext());
        if (!cur.isClosed()) {
            cur.close();
            cur = null;
        }
    } else {
        smsBuilder.append("没有短信!");
    } // end if
    smsBuilder.append("-------End!-------");
} catch (SQLiteException ex) {
    Log.d("SQLiteException in getSmsInPhone", ex.getMessage());
}
return smsBuilder.toString();
}
}
```

步骤03 修改"AndroidManifest.xml"文件,增加android.permission.READ_SMS权限,代码如下。

`<uses-permission android:name="android.permission.READ_SMS"/>`

7.3 简单管理程序设计

本程序可实现简单的通讯录管理功能,主要实现添加联系人,对选定的联系人的信息进行编辑,以及浏览所有联系人的信息。这个管理程序主要运用了Menu、Intent、ContentProvider及SQLite数据库操作。运行应用程序,按Menu菜单时会出现Add Contact菜单,并显示已有联系人的信息,如图7-5所示。点击"Add Contact"按钮时,出现添加联系人的界面,如图7-6所示。点击"Cancel"按钮时，返回前一个界面。当点击"Save"按钮时，显示添加后的信息联系人列表，并在Menu菜单中出现"Edit Contact"（对通讯录进行编辑）菜单项，如图7-7所示。选定一个联系人，长按可以从通讯录中删除该联系人，如图7-8所示。删除后将转向联系人信息浏览界面。

图 7-5　运行开始界面　　图 7-6　添加联系人界面　　图 7-7　联系人信息浏览　　图 7-8　删除选定的联系人

操作步骤如下。

步骤 01 创建项目"SimpleContact"，应用中包含一个名为"ContactEditor"的Activity。

步骤 02 打开"res\values\string.xml"文件，添加对字符串的声明，代码如下。

```xml
<?xml version="1.0" encoding="utf-8"?>
<resources>
    <string name="app_name">SimpleContacts</string>
    <string name="name">Name :</string>
    <string name="mobile">Mobile :</string>
    <string name="email">Email :</string>
    <string name="group">Group :</string>
    <string name="save">Save</string>
    <string name="cancel">Cancel</string>
    <string name="contact_edit">Edit Contact</string>
    <string name="contact_create">Create New Contact</string>
    <string name="error_msg">error in SimpleContacts</string>
    <string name="menu_revert">Revert</string>
    <string name="menu_delete">Delete</string>
    <string name="menu_discard">Discard</string>
    <string name="menu_add">Add Contact</string>
    <string name="menu_edit">Edit Contact</string>
</resources>
```

步骤 03 打开 "res\layout" 文件夹，创建布局文件 "contact_editor.xml"，代码如下。

```xml
<?xml version="1.0" encoding="utf-8"?>
<LinearLayout xmlns:android="http://schemas.android.com/apk/res/android"
    android:orientation="vertical"
    android:layout_width="fill_parent"
    android:layout_height="fill_parent"
    android:background="@drawable/bg">

    <TableRow
        android:id="@+id/TableRow01"
        android:layout_width="fill_parent"
        android:layout_height="wrap_content">
        <TextView
            android:id="@+id/TextView01"
            android:layout_width="wrap_content"
            android:layout_height="wrap_content"
            android:text="@string/name"
            android:textSize="16px"></TextView>
        <EditText
            android:id="@+id/EditText01"
            android:layout_height="wrap_content"
            android:layout_width="fill_parent" ></EditText>
    </TableRow>
    <TableRow
        android:id="@+id/TableRow02"
        android:layout_width="fill_parent"
        android:layout_height="wrap_content">
        <TextView
            android:id="@+id/TextView02"
            android:layout_width="wrap_content"
            android:layout_height="wrap_content"
            android:textSize="16px"
            android:text="@string/mobile"></TextView>
        <EditText
            android:id="@+id/EditText02"
            android:layout_height="wrap_content"
```

```xml
            android:layout_width="fill_parent"></EditText>
    </TableRow>
    <TableRow
        android:id="@+id/TableRow03"
        android:layout_width="fill_parent"
        android:layout_height="wrap_content">
        <TextView
            android:id="@+id/TextView03"
            android:layout_width="wrap_content"
            android:layout_height="wrap_content"
            android:text="@string/email"
            android:textSize="16px"></TextView>
        <EditText
            android:id="@+id/EditText03"
            android:layout_height="wrap_content"
            android:layout_width="fill_parent"></EditText>
    </TableRow>
    <TableRow
        android:id="@+id/TableRow05"
        android:layout_width="fill_parent"
        android:layout_height="wrap_content">
        <Button
            android:id="@+id/Button01"
            android:layout_height="wrap_content"
            android:text="@string/save"
            android:textSize="16px"
            android:layout_width="wrap_content"></Button>
        <Button
            android:id="@+id/Button02"
            android:layout_height="wrap_content"
            android:text="@string/cancel"
            android:textSize="16px"
            android:layout_width="wrap_content"></Button>
    </TableRow>
</LinearLayout>
```

步骤 04 打开"res\layout"文件夹，创建用于显示查询信息的布局文件"contact_list.xml"，代码如下。

```xml
<?xml version="1.0" encoding="utf-8"?>
<LinearLayout xmlns:android="http://schemas.android.com/apk/res/android"
    android:orientation="vertical"
    android:layout_width="fill_parent"
    android:layout_height="fill_parent"
    android:background="@drawable/bg">
<ListView
    android:id="@+id/ListView01"
    android:layout_width="fill_parent"
    android:layout_height="wrap_content"></ListView>
</LinearLayout>
```

步骤 05 打开"res\layout"文件夹，创建用于显示每一项查询信息的布局文件"contact_list_item.xml"，代码如下。

```xml
<?xml version="1.0" encoding="utf-8"?>
<LinearLayout xmlns:android="http://schemas.android.com/apk/res/android"
    android:orientation="vertical"
    android:layout_width="fill_parent"
    android:layout_height="fill_parent">
<TextView xmlns:android="http://schemas.android.com/apk/res/android"
    android:id="@android:id/text1"
    android:layout_width="fill_parent"
    android:layout_height="fill_parent"
    android:textStyle="bold"
    android:textSize="18px"
    android:gravity="center_vertical"
    android:paddingLeft="10px"
    android:singleLine="true"
/>
<TextView xmlns:android="http://schemas.android.com/apk/res/android"
    android:id="@android:id/text2"
    android:layout_width="fill_parent"
    android:layout_height="fill_parent"
    android:textStyle="normal"
    android:textSize="14px"
```

```xml
        android:gravity="center_vertical"
        android:paddingLeft="10px"
        android:singleLine="true"
        />
</LinearLayout>
```

步骤 06 在src文件夹中创建"DBHelper"类,它继承自SQLiteOpenHelper类。

```java
package com.chapt6.contact;
import android.content.Context;
import android.database.sqlite.SQLiteDatabase;
import android.database.sqlite.SQLiteOpenHelper;

public class DBHelper extends SQLiteOpenHelper {
    public static final String DATABASE_NAME = "simplecontacts.db";   //数据库名
    public static final int DATABASE_VERSION = 2;
    public static final String CONTACTS_TABLE = "contacts";   //数据表名
    //创建数据库
    private static final String DATABASE_CREATE = "CREATE TABLE " + CONTACTS_TABLE +" ("
        + ContactColumn._ID+" integer primary key autoincrement,"
        + ContactColumn.NAME+" text,"+ ContactColumn.MOBILE+" text,"
        + ContactColumn.EMAIL+" text,"+ ContactColumn.CREATED+" long,"
        + ContactColumn.MODIFIED+" long);";

    public DBHelper(Context context) {
        super(context, DATABASE_NAME, null, DATABASE_VERSION);
    }

    public void onCreate(SQLiteDatabase db) {
        db.execSQL(DATABASE_CREATE);
    }

    public void onUpgrade(SQLiteDatabase db, int oldVersion, int newVersion) {
        db.execSQL("DROP TABLE IF EXISTS " + CONTACTS_TABLE);
        onCreate(db);
    }
}
```

步骤 07 步骤06中的类"ContactColumn"是为了使用方便而把数据表"contacts"中的列名、列的索引值及其查询字段等字符串合并在一起定义的类。在src文件夹中创建"ContactColumn"类并实现BaseColumns接口,代码如下。

```java
package com.chapt6.contact;
import android.provider.BaseColumns;
public class ContactColumn implements BaseColumns {
    public ContactColumn() {
    }
    // 列名
    public static final String NAME = "name";
    public static final String MOBILE = "mobileNumber";
    public static final String EMAIL = "email";
    public static final String CREATED = "createdDate";
    public static final String MODIFIED = "modifiedDate";
    // 列索引值
    public static final int _ID_COLUMN = 0;
    public static final int NAME_COLUMN = 1;
    public static final int MOBILE_COLUMN = 2;
    public static final int EMAIL_COLUMN = 3;
    public static final int CREATED_COLUMN = 4;
    public static final int MODIFIED_COLUMN = 5;
    // 查询结果
    public static final String[] PROJECTION = {
        _ID,// 0
        NAME,// 1
        MOBILE,// 2
        EMAIL // 3
    };
}
```

步骤 08 在src文件夹中创建类"ContactsProvider",它继承自ContentProvider。在该类中需要实现6个方法,代码如下。

```java
package com.chapt6.contact;
import android.content.ContentProvider;
import android.content.ContentUris;
import android.content.ContentValues;
import android.content.UriMatcher;
```

```java
import android.database.Cursor;
import android.database.SQLException;
import android.database.sqlite.SQLiteDatabase;
import android.database.sqlite.SQLiteQueryBuilder;
import android.net.Uri;
import android.text.TextUtils;
import android.util.Log;

public class ContactsProvider extends ContentProvider {
    private static final String TAG = "ContactsProvider";
    private DBHelper dbHelper;
    private SQLiteDatabase contactsDB;
    public static final String AUTHORITY = "com.chapt6.provider.contact";
    public static final String CONTACTS_TABLE = "contacts";
    public static final Uri CONTENT_URI = Uri.parse("content://" + AUTHORITY + "/contacts");
    public static final int CONTACTS = 1;
    public static final int CONTACT_ID = 2;
    private static final UriMatcher uriMatcher;
    static {
        uriMatcher = new UriMatcher(UriMatcher.NO_MATCH);
        uriMatcher.addURI(AUTHORITY, "contacts", CONTACTS);
        // 单独列
        uriMatcher.addURI(AUTHORITY, "contacts/#", CONTACT_ID);
    }

    public boolean onCreate() {
        dbHelper = new DBHelper(getContext());
        contactsDB = dbHelper.getWritableDatabase();
        return (contactsDB == null) ? false : true;
    }
    // 删除指定数据列
    public int delete(Uri uri, String where, String[] selectionArgs) {
        // TODO Auto-generated method stub
        int count;
        switch (uriMatcher.match(uri)) {
        case CONTACTS:
            count = contactsDB.delete(CONTACTS_TABLE, where, selectionArgs);
```

```java
        break;
    case CONTACT_ID:
        String contactID = uri.getPathSegments().get(1);
        count = contactsDB.delete(CONTACTS_TABLE,
            ContactColumn._ID+ "=" + contactID
            + (!TextUtils.isEmpty(where) ? " AND (" + where + ")" : ""), selectionArgs);
        break;
    default:
        throw new IllegalArgumentException("Unsupported URI: " + uri);
    }
    getContext().getContentResolver().notifyChange(uri, null);
    return count;
}
// Uri类型转换
public String getType(Uri uri) {
    switch (uriMatcher.match(uri)) {
    case CONTACTS:
        return "vnd.android.cursor.dir/contacts";
    case CONTACT_ID:
        return "vnd.android.cursor.item/contacts";
    default:
        throw new IllegalArgumentException("Unsupported URI: " + uri);
    }
}
// 插入数据
public Uri insert(Uri uri, ContentValues initialValues) {
    if (uriMatcher.match(uri) != CONTACTS) {
        throw new IllegalArgumentException("Unknown URI " + uri);
    }
    ContentValues values;
    if (initialValues != null) {
        values = new ContentValues(initialValues);
        Log.e(TAG + "insert", "initialValues is not null");
    } else {
        values = new ContentValues();
    }
    Long now = Long.valueOf(System.currentTimeMillis());
```

```java
// 设置默认值
if (values.containsKey(ContactColumn.CREATED) == false) {
    values.put(ContactColumn.CREATED, now);
}
if (values.containsKey(ContactColumn.MODIFIED) == false) {
    values.put(ContactColumn.MODIFIED, now);
}
if (values.containsKey(ContactColumn.NAME) == false) {
    values.put(ContactColumn.NAME, "");
    Log.e(TAG + "insert", "NAME is null");
}
if (values.containsKey(ContactColumn.MOBILE) == false) {
    values.put(ContactColumn.MOBILE, "");
}
if (values.containsKey(ContactColumn.EMAIL) == false) {
    values.put(ContactColumn.EMAIL, "");
}
Log.e(TAG + "insert", values.toString());
long rowId = contactsDB.insert(CONTACTS_TABLE, null, values);
if (rowId > 0) {
    Uri noteUri = ContentUris.withAppendedId(CONTENT_URI, rowId);
    getContext().getContentResolver().notifyChange(noteUri, null);
    Log.e(TAG + "insert", noteUri.toString());
    return noteUri;
}
throw new SQLException("Failed to insert row into " + uri);
}
// 查询数据
public Cursor query(Uri uri, String[] projection, String selection,
        String[] selectionArgs, String sortOrder) {
    Log.e(TAG + ":query", " in Query");
    SQLiteQueryBuilder qb = new SQLiteQueryBuilder();
    qb.setTables(CONTACTS_TABLE);
    switch (uriMatcher.match(uri)) {
    case CONTACT_ID:
        qb.appendWhere(ContactColumn._ID + "=" + uri.getPathSegments().get(1));
        break;
```

```java
        default:
            break;
        }
        String orderBy;
        if (TextUtils.isEmpty(sortOrder)) {
            orderBy = ContactColumn._ID;
        } else {
            orderBy = sortOrder;
        }
        Cursor c = qb.query(contactsDB, projection, selection, selectionArgs, null, null, orderBy);
        c.setNotificationUri(getContext().getContentResolver(), uri);
        return c;
    }
    // 更新数据库
    public int update(Uri uri, ContentValues values, String where, String[] selectionArgs) {
        int count;
        Log.e(TAG + "update", values.toString());
        Log.e(TAG + "update", uri.toString());
        Log.e(TAG + "update :match", "" + uriMatcher.match(uri));
        switch (uriMatcher.match(uri)) {
        case CONTACTS:
            Log.e(TAG + "update", CONTACTS + "");
            count = contactsDB.update(CONTACTS_TABLE, values, where, selectionArgs);
            break;
        case CONTACT_ID:
            String contactID = uri.getPathSegments().get(1);
            Log.e(TAG + "update", contactID + "");
            count = contactsDB.update(CONTACTS_TABLE,values, ContactColumn._ID
                + "="+ contactID+ (!TextUtils.isEmpty(where) ? " AND (" + where+ ")" : ""), selectionArgs);
            break;
        default:
            throw new IllegalArgumentException("Unsupported URI: " + uri);
        }
        getContext().getContentResolver().notifyChange(uri, null);
        return count;
    }
}
```

步骤 09 修改 "src\ContactEditor.java" 文件，代码如下。

```java
package com.chapt6.contact;
import android.app.Activity;
import android.content.ContentValues;
import android.content.Intent;
import android.database.Cursor;
import android.net.Uri;
import android.os.Bundle;
import android.util.Log;
import android.view.Menu;
import android.view.MenuItem;
import android.view.View;
import android.view.View.OnClickListener;
import android.widget.Button;
import android.widget.EditText;

public class ContactEditor extends Activity {
    private static final String TAG = "ContactEditor";
    private static final int STATE_EDIT = 0;
    private static final int STATE_INSERT = 1;
    private static final int REVERT_ID = Menu.FIRST;
    private static final int DISCARD_ID = Menu.FIRST + 1;
    private static final int DELETE_ID = Menu.FIRST + 2;
    private int mState;
    private Uri mUri;
    private Cursor mCursor;
    private EditText nameText;
    private EditText mPhoneText;
    private EditText emailText;
    private Button saveButton;
    private Button cancelButton;
    private String originalNameText = "";
    private String originalMPhoneText = "";
    private String originalEmailText = "";

    public void onCreate(Bundle savedInstanceState) {
        super.onCreate(savedInstanceState);
```

```java
final Intent intent = getIntent();
final String action = intent.getAction();
Log.e(TAG + ":onCreate", action);
if (Intent.ACTION_EDIT.equals(action)) {
    mState = STATE_EDIT;
    mUri = intent.getData();
} else if (Intent.ACTION_INSERT.equals(action)) {
    mState = STATE_INSERT;
    mUri = getContentResolver().insert(intent.getData(), null);
    if (mUri == null) {
        Log.e(TAG + ":onCreate", "Failed to insert new Contact into " + getIntent().getData());
        finish();
        return;
    }
    setResult(RESULT_OK, (new Intent()).setAction(mUri.toString()));
} else {
    Log.e(TAG + ":onCreate", " unknown action");
    finish();
    return;
}
setContentView(R.layout.contact_editor);
nameText = (EditText) findViewById(R.id.EditText01);
mPhoneText = (EditText) findViewById(R.id.EditText02);
emailText = (EditText) findViewById(R.id.EditText03);
saveButton = (Button) findViewById(R.id.Button01);
cancelButton = (Button) findViewById(R.id.Button02);
saveButton.setOnClickListener(new OnClickListener() {
    public void onClick(View v) {
        String text = nameText.getText().toString();
        if (text.length() == 0) {
            setResult(RESULT_CANCELED);
            deleteContact();
            finish();
        } else {
            updateContact();
        }
    }
}
```

```java
    });
    cancelButton.setOnClickListener(new OnClickListener() {
      public void onClick(View v) {
        if (mState == STATE_INSERT) {
          setResult(RESULT_CANCELED);
          deleteContact();
          finish();
        } else {
          backupContact();
        }
      }
    });
    Log.e(TAG + ":onCreate", mUri.toString());
    // 获得并保存原始联系人信息
    mCursor = managedQuery(mUri, ContactColumn.PROJECTION, null, null, null);
    mCursor.moveToFirst();
    originalNameText = mCursor.getString(ContactColumn.NAME_COLUMN);
    originalMPhoneText = mCursor.getString(ContactColumn.MOBILE_COLUMN);
    originalEmailText = mCursor.getString(ContactColumn.EMAIL_COLUMN);
    Log.e(TAG, "end of onCreate()");
}

protected void onResume() {
    super.onResume();
    if (mCursor != null) {
      Log.e(TAG + ":onResume", "count:" + mCursor.getColumnCount());
      // 读取并显示联系人信息
      mCursor.moveToFirst();
      if (mState == STATE_EDIT) {
        setTitle(getText(R.string.contact_edit));
      } else if (mState == STATE_INSERT) {
        setTitle(getText(R.string.contact_create));
      }
      String name = mCursor.getString(ContactColumn.NAME_COLUMN);
      String mPhone = mCursor.getString(ContactColumn.MOBILE_COLUMN);
      String email = mCursor.getString(ContactColumn.EMAIL_COLUMN);
      nameText.setText(name);
```

```java
        mPhoneText.setText(mPhone);
        emailText.setText(email);
    } else {
        setTitle(getText(R.string.error_msg));
    }
}

protected void onPause() {
    super.onPause();
    if (mCursor != null) {
        String text = nameText.getText().toString();
        if (text.length() == 0) {
            Log.e(TAG + ":onPause", "nameText is null ");
            setResult(RESULT_CANCELED);
            deleteContact();
            // 更新信息
        } else {
        ContentValues values = new ContentValues();
        values.put(ContactColumn.NAME, nameText.getText().toString());
        values.put(ContactColumn.MOBILE, mPhoneText.getText()    .toString());
        values.put(ContactColumn.EMAIL, emailText.getText().toString());
        getContentResolver().update(mUri, values, null, null);
        }
    }
}

public boolean onCreateOptionsMenu(Menu menu) {
    if (mState == STATE_EDIT) {
        menu.add(0, REVERT_ID, 1, R.string.menu_revert).setShortcut('0', 'r')
            .setIcon(android.R.drawable.ic_menu_revert);
        menu.add(0, DELETE_ID, 2, R.string.menu_delete).setShortcut('0', 'd')
            .setIcon(android.R.drawable.ic_menu_delete);
    } else {
        menu.add(0, DISCARD_ID, 3, R.string.menu_discard).setShortcut('0', 'd')
            .setIcon(android.R.drawable.ic_menu_delete);
    }
    return true;
```

```java
    }

    public boolean onOptionsItemSelected(MenuItem item) {
        switch (item.getItemId()) {
        case DELETE_ID:
            deleteContact();
            finish();
            break;
        case DISCARD_ID:
            cancelContact();
            break;
        case REVERT_ID:
            backupContact();
            break;
        }
        return super.onOptionsItemSelected(item);
    }

    // 删除联系人信息
    private void deleteContact() {
        if (mCursor != null) {
            mCursor.close();
            mCursor = null;
            getContentResolver().delete(mUri, null, null);
            nameText.setText("");
        }
    }

    // 丢弃信息
    private void cancelContact() {
        if (mCursor != null) {
            deleteContact();
        }
        setResult(RESULT_CANCELED);
        finish();
    }
```

```java
// 更新并变更的信息
private void updateContact() {
    if (mCursor != null) {
        mCursor.close();
        mCursor = null;
        ContentValues values = new ContentValues();
        values.put(ContactColumn.NAME, nameText.getText().toString());
        values.put(ContactColumn.MOBILE, mPhoneText.getText().toString());
        values.put(ContactColumn.EMAIL, emailText.getText().toString());
        getContentResolver().update(mUri, values, null, null);
    }
    setResult(RESULT_CANCELED);
    finish();
}

// 取消操作，回退到最初的信息
private void backupContact() {
    if (mCursor != null) {
        mCursor.close();
        mCursor = null;
        ContentValues values = new ContentValues();
        values.put(ContactColumn.NAME, this.originalNameText);
        values.put(ContactColumn.MOBILE, this.originalMPhoneText);
        values.put(ContactColumn.EMAIL, this.originalEmailText);
        getContentResolver().update(mUri, values, null, null);
    }
    setResult(RESULT_CANCELED);
    finish();
}
```

步骤 10 在src文件夹中创建"Contacts"类，它继承自ListActivity，该类用于显示联系人的信息和编辑联系人的界面，代码如下。

```java
package com.chapt6.contact;
import android.app.ListActivity;
import android.content.ComponentName;
import android.content.ContentUris;
```

```java
import android.content.Intent;
import android.database.Cursor;
import android.net.Uri;
import android.os.Bundle;
import android.util.Log;
import android.view.ContextMenu;
import android.view.Menu;
import android.view.MenuItem;
import android.view.View;
import android.view.ContextMenu.ContextMenuInfo;
import android.widget.AdapterView;
import android.widget.ListView;
import android.widget.SimpleCursorAdapter;

public class Contacts extends ListActivity {
    private static final String TAG = "Contacts";
    private static final int AddContact_ID = Menu.FIRST;
    private static final int EditContact_ID = Menu.FIRST+1;

    public void onCreate(Bundle savedInstanceState) {
    super.onCreate(savedInstanceState);
    setDefaultKeyMode(DEFAULT_KEYS_SHORTCUT);
    Intent intent = getIntent();
    if (intent.getData() == null) {
       intent.setData(ContactsProvider.CONTENT_URI);
    }
    getListView().setOnCreateContextMenuListener(this);
    Cursor cursor = managedQuery(getIntent().getData(), ContactColumn.PROJECTION, null, null,null);
    //注册每个列表的表示形式：姓名＋手机号码
    SimpleCursorAdapter adapter = new SimpleCursorAdapter(this, R.layout.contact_list_item, cursor,
       new String[] { ContactColumn.NAME,ContactColumn.MOBILE }, new int[] { android.R.id.text1,
android.R.id.text2 });
    setListAdapter(adapter);
    Log.e(TAG+"onCreate"," is ok");
    }

    public boolean onCreateOptionsMenu(Menu menu) {
```

```java
    super.onCreateOptionsMenu(menu);
    menu.add(0, AddContact_ID, 0, R.string.menu_add).setShortcut('3', 'a')
        .setIcon(android.R.drawable.ic_menu_add);
    Intent intent = new Intent(null, getIntent().getData());
    intent.addCategory(Intent.CATEGORY_ALTERNATIVE);
    menu.addIntentOptions(Menu.CATEGORY_ALTERNATIVE, 0, 0, new ComponentName(this,
                    Contacts.class), null, intent, 0, null);
    return true;
}

public boolean onPrepareOptionsMenu(Menu menu) {
    super.onPrepareOptionsMenu(menu);
    final boolean haveItems = getListAdapter().getCount() > 0;
    if (haveItems) {
        Uri uri = ContentUris.withAppendedId(getIntent().getData(), getSelectedItemId());
        Intent[] specifics = new Intent[1];
        specifics[0] = new Intent(Intent.ACTION_EDIT, uri);
        MenuItem[] items = new MenuItem[1];
        Intent intent = new Intent(null, uri);
        intent.addCategory(Intent.CATEGORY_ALTERNATIVE);
        menu.addIntentOptions(Menu.CATEGORY_ALTERNATIVE, 0, 0, null, specifics, intent, 0,items);
        if (items[0] != null) {
            items[0].setShortcut('1', 'e');
        }
    } else {
        menu.removeGroup(Menu.CATEGORY_ALTERNATIVE);
    }
    return true;
}

public boolean onOptionsItemSelected(MenuItem item) {
    switch (item.getItemId()) {
    case AddContact_ID:
        //添加联系人
        startActivity(new Intent(Intent.ACTION_INSERT, getIntent().getData()));
        return true;
    }
```

```java
        return super.onOptionsItemSelected(item);
    }

    public void onCreateContextMenu(ContextMenu menu, View view, ContextMenuInfo menuInfo) {
        AdapterView.AdapterContextMenuInfo info;
        try {
            info = (AdapterView.AdapterContextMenuInfo) menuInfo;
        } catch (ClassCastException e) {
            return;
        }
        Cursor cursor = (Cursor) getListAdapter().getItem(info.position);
        if (cursor == null) {
            return;
        }
        menu.setHeaderTitle(cursor.getString(1));
        menu.add(0, EditContact_ID, 0, R.string.menu_delete);
    }

    public boolean onContextItemSelected(MenuItem item) {
        AdapterView.AdapterContextMenuInfo info;
        try {
            info = (AdapterView.AdapterContextMenuInfo) item.getMenuInfo();
        } catch (ClassCastException e) {
            return false;
        }
        switch (item.getItemId()) {
            case EditContact_ID: {
                Uri noteUri = ContentUris.withAppendedId(getIntent().getData(), info.id);
                getContentResolver().delete(noteUri, null, null);
                return true;
            }
        }
        return false;
    }

    @Override
    protected void onListItemClick(ListView l, View v, int position, long id) {
```

```
        Uri uri = ContentUris.withAppendedId(getIntent().getData(), id);
        String action = getIntent().getAction();
        if (Intent.ACTION_PICK.equals(action) || Intent.ACTION_GET_CONTENT.equals(action)) {
            setResult(RESULT_OK, new Intent().setData(uri));
        } else {
            //编辑联系人
            startActivity(new Intent(Intent.ACTION_EDIT, uri));
        }
    }
}
```

步骤 11 修改"AndroidManifest.xml"文件，对定义的Provider和Activity进行注册，代码如下。

```xml
<?xml version="1.0" encoding="utf-8"?>
<manifest xmlns:android="http://schemas.android.com/apk/res/android"
    package="com.chapt6.contact" android:versionCode="1"
    android:versionName="1.0.0">
    <application android:icon="@drawable/icon" android:label="@string/app_name">
        <provider android:name="ContactsProvider"
            android:authorities="com.chapt6.provider.contact" />
        <activity android:name=".Contacts" android:label="@string/app_name">
            <intent-filter>
                <action android:name="android.intent.action.MAIN" />
                <category android:name="android.intent.category.LAUNCHER" />
            </intent-filter>
            <intent-filter>
                <action android:name="android.intent.action.VIEW" />
                <action android:name="android.intent.action.EDIT" />
                <action android:name="android.intent.action.PICK" />
                <category android:name="android.intent.category.DEFAULT" />
                <data android:mimeType="vnd.android.cursor.dir/contacts" />
            </intent-filter>
            <intent-filter>
                <action android:name="android.intent.action.GET_CONTENT" />
                <category android:name="android.intent.category.DEFAULT" />
                <data android:mimeType="vnd.android.cursor.item/contacts" />
            </intent-filter>
        </activity>
```

```xml
<activity android:name=".ContactEditor" android:theme="@android:style/Theme.Light"
    android:label="ContactEditor">
    <intent-filter android:label="@string/menu_edit">
        <action android:name="android.intent.action.VIEW" />
        <action android:name="android.intent.action.EDIT" />
        <action android:name="com.android.notepad.action.EDIT_NOTE" />
        <category android:name="android.intent.category.DEFAULT" />
        <data android:mimeType="vnd.android.cursor.item/contacts" />
    </intent-filter>
    <intent-filter>
        <action android:name="android.intent.action.INSERT" />
        <category android:name="android.intent.category.DEFAULT" />
        <data android:mimeType="vnd.android.cursor.dir/contacts" />
    </intent-filter>
</activity>
</application>
</manifest>
```

课后作业

一、填空题

1. 一个Android应用主要由4种组件组成，分别为_____、_____、_____、_____。

2. 理解Intent的关键之一是理解Intent的两种基本用法：一种是_____，即在构造Intent对象时就指定接收者；另一种是_____，即Intent的发送者在构造Intent对象时，并不知道也不关心接收者是谁，这有利于降低发送者和接收者之间的耦合。

3. 通常Intent的调用分为_____和_____两种。

4. Intent除了定位目标组件外，另外一个职责就是_____。Intent之间传递数据一般有两种常用的方法：一种是_____，另一种是_____。

二、选择题

1. Intent的属性中，(　　)属性是指定Intent的目标组件的类名称的。

　　A. Type

　　B. Data

　　C. Category

　　D. Action

2. （　　）不是Intent组件的常用方式。

　　A. 调用拨号程序

　　B. 通过浏览器打开网页

　　C. 编辑电子邮件

　　D. 显示地图与路径规划

3. 在一个Activity中，可能会使用（　　）方法打开多个不同的Activity处理不同的业务。

　　A. setResult(int resultCode, Intent data)

　　B. onActivityResult(int requestCode, int resultCode, Intent data)

　　C. getIntent()

　　D. startActivityForResult()

4. 使用Intent启动不同组件的方法，（　　）是启动Activity的方法。

　　A. MainActivity

　　B. ACTION_SYNC

　　C. Context.startActivity(Intent intent)

　　D. OtherActivity

三、操作题

利用Android Studio开发环境创建一个新的Android应用程序，实现启动Android自带的打电话功能Dialer程序的功能。（提示：参照7.1节中的内容进行练习。）

第 8 章

多媒体开发和电话API

内容概要

本章主要介绍Android平台的多媒体开发方面的知识，多媒体主要包括音频和视频。本章主要介绍音频和视频的播放、音频和视频的录制，以及如何拨打电话和发送、接收短信等内容。

数字资源

【本章案例文件】:"案例文件\第8章"目录下

配套资源
入门精讲
项目实战
日志记录

8.1 多媒体开发

智能手机除了常规的拨打电话功能之外，还可以浏览互联网、听音乐、看视频，甚至观看在线电影。Android平台提供了强大的多媒体功能，它支持相当广泛的音频和视频格式。本节将先介绍Android平台支持的常用音频、视频格式，然后介绍播放音频和视频的方法，最后再介绍如何录制音频和视频。

8.1.1 常见的多媒体格式

Android支持多种音频格式和编解码器，下面是几种常用的音频格式。

1. AMR

AMR是指自适应多速率编解码器，包括AMR窄带（AMR-NB）和AMR宽带（AMR-WB）两种，文件扩展名是".3gp"或".amr"。AMR是基本音频编解码标准，主要应用于手机的语音通话，并得到了手机厂商的广泛支持。该编码标准适用于简单的语音编码，不适用处理更复杂的音频数据，如音乐等。

2. AAC

AAC的全称是Advanced Audio Coding，它是一种专为声音数据设计的文件压缩格式。AAC与MP3不同，它采用了全新的算法进行编码，更加高效，性价比也更高。利用AAC格式，可使声音质量没有明显降低的前提下文件更小。Android除了支持AAC之外，还支持新添加到AAC规范中的高效AAC（High Efficiency AAC）格式。

3. MP3

MP3是一种音频压缩技术，其全称是动态影像专家压缩标准音频层面3（Moving Picture Experts Group Audio Layer III），简称为MP3。MP3可大幅度降低音频数据量，利用MPEG Audio Layer III技术，可以将音乐以1∶10甚至1∶12的压缩率压缩成容量较小的文件。对于大多数用户来说，压缩后的音频与最初不压缩的音频相比，音质没有明显的下降。用MP3形式存储的音乐就称为MP3音乐，能播放MP3音乐的机器就叫MP3播放器。MP3是目前互联网上使用最广泛的音频编解码器之一。

4. OGG

OGG的全称是OGG Vorbis，是一种新的音频压缩格式，类似于MP3的音乐格式，但不同的是，它是完全免费、开放和没有专利限制的。OGG Vorbis格式有一个特点是支持多声道。OGG Vorbis格式的文件扩展名是".ogg"。这种文件的设计格式非常先进，创建的OGG文件可以在未来的任何播放器上播放，这种文件格式可以不断对文件大小和音质进行改良，且不影响旧有的编码器或播放器。

除了音频之外，Android还支持多种视频格式和编解码器，并且支持的类型还在不断增加，下面介绍两种常用的视频编码标准。

1. H.263

H.263是由ITU-T(国际电联电信标准化部门)制定的低码率视频编码标准,主要用于低延迟和低比特率的视频会议应用中。MPEG-4".mp4"和3GP".3gp"文件都支持H.263编码的视频。

2. H.264

H.264也称为MPEG-4 part 10或AVC(Advanced Video Coding,高级视频编码)。它是视频编解码器的最新标准,在软件和硬件方面获得广泛支持。Silverlight、Flash、iPhone/iPod以及蓝光等设备都支持H.264。Android以MPEG-4容器格式(文件扩展名为".mp4")支持H.264编码的视频。

■8.1.2 播放音频

在Android中使用MediaPlayer类播放音频。MediaPlayer类是Android SDK提供的播放音频和视频的功能类,使用它可以建立功能更加完善的音频播放应用。

播放音频的最简单方式是播放与应用程序本身一起打包的音频文件,此时音频文件放置在应用程序的原始资源中。

在项目的res文件夹中创建一个新文件夹,命名为"raw",把音频文件放置在该文件夹中,ADT将会自动更新"R.java"文件(位于"gen"文件夹中),为该音频文件生成资源id,使用R.raw.file_name_without_extension属性可访问该音频文件。

播放与应用程序一起打包的音频文件非常简单。使用MediaPlayer类的静态方法create实例化一个MediaPlayer对象,传入参数this和音频文件的资源id即可,代码如下。

```
MediaPlayer mediaPlayer = MediaPlayer.create(this, R.raw.audio_file_name_without_extension);
```

由于调用MediaPlayer类的静态方法create创建MediaPlayer对象成功后,系统会自动调用prepare方法,不需要再手动调用,MediaPlayer对象已经处于Prepared状态,因此只需调用MediaPlayer对象的start方法即可播放该音频文件,代码如下。

```
mediaPlayer.start();
```

MediaPlayer类内部维护一个状态机来管理播放音频和视频中的各种状态,该状态机如图8-1所示(引用自Android developers网站的"API参考"栏目),它描述了音频和视频播放过程中的各种状态和每个状态下可以调用的方法。

这里需要着重强调的是MediaPlayer对象是基于状态的。在写代码时,必须始终注意MediaPlayer对象所处的状态,因为MediaPlayer对象的方法都有其可以正常执行的有效状态。如果在一个错误的状态执行一个方法时,系统会抛出异常或者产生不可预料的行为。

MediaPlayer的状态机图中明确指出了哪些方法可以把MediaPlayer对象从一个状态迁移到另一个状态。例如,当新建一个MediaPlayer对象时(使用new操作),它处于Idle状态。在Idle状态时,通过调用setDataSource方法初始化MediaPlayer对象,迁移到Initialized状态。之后,使用prepare或prepareAsync方法准备MediaPlayer对象。当MediaPlayer对象准备就绪时，它进入Prepared状态，这时可以调用start方法来播放媒体文件。此时，MediaPlayer对象处于Started状态。由图8-1可以看出，通过调用start、pause和seekTo方法，可以使得MediaPlayer在Started、Paused和PlaybackCompleted 3个状态之间进行迁移。当调用stop方法停止播放后，就不能再调用start方法播放媒体文件了，除非再次调用prepare方法准备MediaPlayer对象。

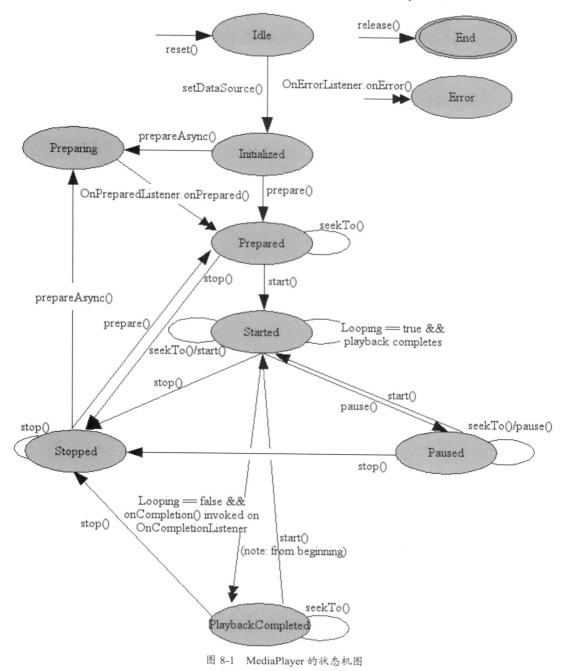

图 8-1 MediaPlayer 的状态机图

再次强调：编写媒体播放代码时，一定注意MediaPlayer对象的状态，因为在不同的状态可调用的方法不同，使用错误的调用方法会造成错误或出现不可预料的行为。

使用MediaPlayer类播放时，需要注意不要在应用程序的UI主线程中准备MediaPlayer。调用prepare方法通常会花费一定的时间，因为它需要获取并解码媒体数据。只要是任何需要花费一定时间执行的方法，都不要在应用程序的UI主线程中调用。如果那样做的话，系统将会挂起UI主线程，直到该方法执行完毕。这是一个非常差的用户体验，会造成应用程序不响应（Application Not Responding）错误。即使是很短的媒体数据，加载可能会很快，但是只要超过100 ms的操作都会造成UI界面发生明显的停顿，会让用户感觉应用程序响应很慢。

为了避免挂起UI主线程，可创建一个新的线程准备MediaPlayer，当准备就绪时，通知UI主线程。其实，无须编写新的线程逻辑，MediaPlayer对象本身提供了一个非常方便的方法可以完成该任务，即prepareAsync方法。该方法在后台准备媒体数据，准备就绪后立刻返回。当媒体数据准备好后，MediaPlayer.OnPreparedListener接口的onPrepared回调方法会自动被调用，以便继续后续处理。通过MediaPlayer的setOnPreparedListener方法可以注册监听器。

MediaPlayer有可能会耗尽系统资源，因此应该采取措施避免MediaPlayer一直占用系统资源。当使用MediaPlayer播放完毕时，应该调用release方法确保任何分配给它的系统资源都被合适地释放。例如，如果应用的Activity收到onStop回调方法的调用，则必须在其中释放MediaPlayer，因为当该Activity不与用户交互时，MediaPlayer还占用系统资源已经没有意义，除非它正在后台播放。当Activity还继续运行或重启动时，则需要创建一个新的MediaPlayer，并再次准备它，之后才能继续播放。

```
mediaPlayer.release();
mediaPlayer = null;
```

【案例8-1】：演示如何播放MP3文件。

一般来说，播放MP3音频文件需要遵循如下步骤。

（1）把将要播放的MP3文件放到Android项目的"res\raw"目录下，也可以使用Uri访问网络上的音频文件。

（2）通过调用MediaPlayer.create()方法创建MediaPlayer的实例对象，并引用该MP3文件。

（3）调用MediaPlayer的prepare和start方法。

操作步骤如下。

步骤 01 创建一个新的Android项目"MediaPlayerExample"，创建一个Activity，命名为"MediaPlayerActivity"。

步骤 02 在项目的res文件夹下创建一个新目录"raw"，把要播放的MP3文件放到该目录下，然后创建一个简单的按钮，用于控制MP3的播放。布局XML代码如下。

```xml
<?xml version="1.0" encoding="utf-8"?>
<LinearLayout xmlns:android="http://schemas.android.com/apk/res/android"
    android:orientation="vertical"
    android:layout_width="fill_parent"
    android:layout_height="fill_parent"
    >
    <TextView
    android:layout_width="fill_parent"
    android:layout_height="wrap_content"
    android:text="Simple Media Player"
    />
    <Button android:id="@+id/playsong"
    android:layout_width="fill_parent"
    android:layout_height="wrap_content"
    android:text="an MP3 audio file"
    />
</LinearLayout>
```

步骤 03 修改 "MediaPlayerActivity.java" 文件的代码，代码如下。

```java
public class MediaPlayerActivity extends Activity {
    public void onCreate(Bundle icicle) {
        super.onCreate(icicle);
        setContentView(R.layout.main);
        Button mybutton = (Button) findViewById(R.id.playsong);
        mybutton.setOnClickListener(new Button.OnClickListener() {
            public void onClick(View v) {
                MediaPlayer mp = MediaPlayer.create(MediaPlayerActivity.this, R.raw.halotheme);
                mp.start();
                mp.setOnCompletionListener(new OnCompletionListener(){
                    public void onCompletion(MediaPlayer arg0) {
                    }
                });
            }
        });
    }
}
```

播放MP3文件很简单，只需使用布局文件创建的视图View显示界面，映射资源id为"playsong"的资源到"mybutton"对象；然后绑定到setOnClickListener方法，在该监听器方法中已经使用create(Context context, int resourceid)方法创建了MediaPlayer类的实例；最后设置setOnCompletionListener方法，当播放完成后，执行某些资源释放的操作。当前的例子没有做任何事情，如果想要改变按钮状态，想通知用户音乐播放完毕，或询问用户是否要播放另一首歌曲时，可以在该方法中做相应的设置。

8.1.3 播放视频

Android平台提供一个VideoView widget来处理视频画面的显示，可以把它放到任何界面布局管理器中。此外，Android还提供了大量的显示选项，包括缩放和着色。

【案例8-2】：演示如何播放视频文件。

操作步骤如下。

步骤 01 创建界面布局，该布局界面中增加了一个VideoView widget，它提供视频画面显示的功能，并且包含停止、播放、快进、快退等一些功能按钮。布局文件代码如下。

```xml
<?xml version="1.0" encoding="utf-8"?>
<LinearLayout xmlns:android="http://schemas.android.com/apk/res/android"
    android:orientation="vertical"
    android:layout_width="fill_parent"
    android:layout_height="fill_parent"
    >
    <VideoView android:id="@+id/video"
    android:layout_width="320px"
    android:layout_height="240px"
    />
</LinearLayout>
```

步骤 02 编写一个类文件"SimpleVideo"，用于实现播放视频的功能，代码如下。

```java
public class SimpleVideo extends Activity {
    private VideoView myVideo;
    private MediaController mc;
    public void onCreate(Bundle icicle) {
        super.onCreate(icicle);
        getWindow().setFormat(PixelFormat.TRANSLUCENT);
        setContentView(R.layout.main);
        myVideo = (VideoView) findViewById(R.id.video);
        File pathToTest= new File (Environment.getExternalFileDirectory(),"test.mp4");
```

```
        myVideo.setVideoPath(pathToTest);
        mc = new MediaController(this);
        mc.setMediaPlayer(myVideo);
        myVideo.setMediaController(mc);
        myVideo.requestFocus();
    }
}
```

在这个播放视频的类文件中，首先创建一个TRANSLUCENT窗口，这个窗口对于SurfaceView来说是必须的；接着把VideoView作为播放视频的容器，并且使用setVideoPath方法设置将要播放的视频文件的路径；最后，创建MediaController类的实例对象，使用setMediaController方法向VideoView注册回调，以便完成视频播放后通知它。

运行该Android应用之前，可以使用ADB工具推送视频文件到Android设备上。当视频文件位于设备的SD卡上时，运行该应用并且触摸屏幕，控制按钮将会出现。

VideoView和MediaPlayer都简化了视频的播放，但要注意的是，模拟器和真实设备在处理较大的视频文件时的反应会有所不同。

8.1.4 录制音频

在Android中不仅播放音频、视频是常用功能，录制音频、视频也是经常用到的操作之一。本节主要介绍音频的录制。

录制音频大致遵循以下9个步骤。

步骤 01 创建android.media.MediaRecorder对象实例，之后所有的工作都将围绕该对象展开。

```
MediaRecorder mRecorder = new MediaRecorder();
```

步骤 02 设置音频录制时采用的音频源（audio source），通过调用前面创建的MediaRecorder对象的MediaRecorder.setAudioSource()方法完成。MediaRecorder.AudioSource是MediaRecorder的内部类，它主要定义智能手机常用的音频源资源。MediaRecorder.AudioSource.MIC是最常用的音频源，代表手机的麦克风外设。除了MediaRecorder.AudioSource.MIC以外，还提供VOICE_CALL、VOICE_DOWNLINK、VOICE_UPLINK音频源，可以通过这些音频源进行语音通话的录制。例如：

```
mRecorder.setAudioSource(MediaRecorder.AudioSource.MIC);
```

步骤 03 设置输出音频数据的存储文件格式。这里，通过调用MediaRecorder.setOutputFormat()方法完成。Android平台支持的音频文件格式由MediaRecorder.OutputFormat内部类定义。Android支持的主要文件格式有以下几种。

- **MediaRecorder.OutputFormat.AMR_NB**：该常量表示输出文件是AMR-NB格式的语音

文件，主要对人声进行编码。
- **MediaRecorder.OutputFormat.MPEG_4**：该常量表示输出文件是MPEG-4格式的多媒体文件，其中可能同时包含音频和视频信息。
- **MediaRecorder.OutputFormat.THREE_GPP**：该常量表示输出文件是3GPP格式的文件，其中可能同时包含音频和视频信息。

例如：

mRecorder.setOutputFormat(MediaRecorder.OutputFormat.AMR_NB);

步骤04 指定输出音频数据存放的文件。这里通过调用MediaRecorder. setOutputFile()方法完成。该方法可以接收两种参数：文件描述符（FileDescriptor）和文件路径字符串。

FileDescriptor fd = …;
mRecorder.setOutputFile(fd);

或者

String mFileName = …;
mRecorder.setOutputFile(mFileName);

步骤05 设置音频编码器。通过调用MediaRecorder.setAudioEncoder()方法完成。Android平台支持的音频编码器由MediaRecorder.AudioEncoder内部类定义。最常用的音频编码器是MediaRecorder.AudioEncoder.AMR_NB。AMR_NB是自适应多速率窄带音频编解码器。该编解码器主要针对语音数据进行优化，使其不适应语音之外的其他音频信息。例如：

mRecorder.setAudioEncoder(MediaRecorder.AudioEncoder.AMR_NB);

步骤06 调用MediaRecorder.prepare()方法，准备工作就绪，可以开始录制音频。例如：

mRecorder.prepare();

步骤07 调用MediaRecorder.start()方法，开始录制。例如：

mRecorder.start();

步骤08 录制完成之后，需要调用MediaRecorder.stop()方法停止录制。例如：

mRecorder.stop();

步骤09 调用MediaRecorder.release()方法，释放所占用的资源。到这里，整个录制过程就完成了。

【案例8-3】：演示如何录制音频文件。

操作步骤如下。

步骤 01 创建一个Android项目"SoundRecordingExample"。

步骤 02 编辑"AndroidManifest.xml"文件，添加如下代码。

```xml
<uses-permission android:name="android.permission.RECORD_AUDIO" />
<uses-permission android:name="android.permission.WRITE_EXTERNAL_STORAGE"/>
```

上面的代码设置了应用程序的权限，允许应用程序录制音频文件并播放它们。

步骤 03 创建类"SoundRecordingDemo"，代码如下。

```java
public class SoundRecordingDemo extends Activity {
    MediaRecorder mRecorder;
    File mSampleFile = null;
    static final String SAMPLE_PREFIX = "recording";
    static final String SAMPLE_EXTENSION = ".3gpp";
    private static final String OUTPUT_FILE = "/sdcard/audiooutput.3gpp";
    private static final String TAG = "SoundRecordingExample";

    public void onCreate(Bundle savedInstanceState) {
        super.onCreate(savedInstanceState);
        setContentView(R.layout.main);
        this.mRecorder = new MediaRecorder();
        Button startRecording = (Button) findViewById(R.id.startrecording);
        Button stopRecording = (Button) findViewById(R.id.stoprecording);
        startRecording.setOnClickListener(new View.OnClickListener() {
            public void onClick(View v) {
                startRecording();
            }
        });
        stopRecording.setOnClickListener(new View.OnClickListener() {
            public void onClick(View v) {
                stopRecording();
                addToDB();
            }
        });
    }

    protected void addToDB() {
        ContentValues values = new ContentValues(3);
        long current = System.currentTimeMillis();
```

```java
        values.put(MediaColumns.TITLE, "test_audio");
        values.put(MediaColumns.DATE_ADDED, (int) (current / 1000));
        values.put(MediaColumns.MIME_TYPE, "audio/3gpp");
        values.put(MediaColumns.DATA, OUTPUT_FILE);
        ContentResolver contentResolver = getContentResolver();
        Uri base = MediaStore.Audio.Media.EXTERNAL_CONTENT_URI;
        Uri newUri = contentResolver.insert(base, values);
        sendBroadcast(new Intent(Intent.ACTION_MEDIA_SCANNER_SCAN_FILE, newUri)); }

    protected void startRecording() {
        this.mRecorder = new MediaRecorder();
        this.mRecorder.setAudioSource(MediaRecorder.AudioSource.MIC);
        this.mRecorder.setOutputFormat(MediaRecorder.OutputFormat.THREE_GPP);
        this.mRecorder.setAudioEncoder(MediaRecorder.AudioEncoder.AMR_NB);
        this.mRecorder.setOutputFile(OUTPUT_FILE);
        try {
            this.mRecorder.prepare();
        } catch (IllegalStateException e1) {
            // TODO Auto-generated catch block
            e1.printStackTrace();
        } catch (IOException e1) {
            // TODO Auto-generated catch block
            e1.printStackTrace();
        }
        this.mRecorder.start();
        if (this.mSampleFile == null) {
            File sampleDir = Environment.getExternalStorageDirectory();
            try {
                this.mSampleFile = File.createTempFile(SoundRecordingDemo.SAMPLE_PREFIX, SoundRecordingDemo.SAMPLE_EXTENSION, sampleDir);
            } catch (IOException e) {
                Log.e(SoundRecordingDemo.TAG, "sdcard access error");
                return;
            }
        }
    }
```

```
    protected void stopRecording() {
        this.mRecorder.stop();
        this.mRecorder.release();
    }
}
```

代码的第一部分中包括创建按钮和设置监听器，监听开始录制和停止录制事件，还有一个重要的方法addToDB，该方法用于设置音频文件的元数据信息，包括标题、日期和文件类型；然后调用Intent的ACTION_MEDIA_SCANNER_SCAN_FILE通知应用程序，一个新的音频文件已经创建。使用该Intent允许查询播放列表中的新的音频文件并播放它们。

接下来是startRecording方法，该方法首先创建一个新的MediaRecorder实例，然后设置音频源和麦克风，设置输出格式为THREE_GPP，设置音频编码器格式为AMR_NB，设置输出文件路径；接着调用prepare和start方法开始录制音频；最后，创建stopRecording方法，调用stop和release方法停止MediaRecorder录制音频并释放资源。

录制音频的所有准备工作就绪后，在模拟器中运行该应用程序。点击开始录制按钮，几秒过后，点击停止录制按钮。打开DDMS，可以在"sdcard"目录下看到录制好的文件。可以使用设备的媒体播放器或浏览器查看并播放该录制的文件。

8.1.5 录制视频

不同于录制音频，录制视频时，Android要求在录制之前必须先预览视频。这可能对某些应用有些不合适，但这是Android 2.2及其以上版本的强制要求。

正如录制音频一样，录制视频之前，必须设置一些权限，包括RECORD_VIDEO、CAMERA、RECORD_AUDIO和WRITE_EXTERNAL_STORAGE。

【案例8-4】：演示如何录制视频。

操作步骤如下。

步骤01 创建一个Android项目"VideoCamExample"，然后编辑"AndroidManifest.xml"文件，代码如下。

```xml
<?xml version="1.0" encoding="utf-8"?>
<manifest xmlns:android="http://schemas.android.com/apk/res/android"
    package="com.selfteaching.VideoCamExample"
    android:versionCode="1"
    android:versionName="1.0">
    <application android:icon="@drawable/icon" android:label="@string/app_name">
        <activity android:name=".VideoCamExample" android:label="@string/app_name">
            <intent-filter>
                <action android:name="android.intent.action.MAIN" />
```

```xml
            <category android:name="android.intent.category.LAUNCHER" />
        </intent-filter>
    </activity>
</application>
<uses-permission android:name="android.permission.CAMERA"></uses-permission>
<uses-permission android:name="android.permission.RECORD_AUDIO"></uses-permission>
<uses-permission android:name="android.permission.RECORD_VIDEO"></uses-permission>
<uses-permission android:name="android.permission.WRITE_EXTERNAL_STORAGE" />
<uses-feature android:name="android.hardware.camera" />
</manifest>
```

该文件最后使用了uses-feature语句，目的是指明该程序的运行依赖于哪些软硬件资源，本例中指明需要使用摄像头。

步骤 02 为应用创建一个简单的布局，包含预览区域、开始、停止、暂停和播放按钮。布局的XML文件代码如下。

```xml
<?xml version="1.0" encoding="utf-8"?>
<RelativeLayout xmlns:android="http://schemas.android.com/apk/res/android"
    android:orientation="vertical"
    android:layout_width="fill_parent"
    android:layout_height="fill_parent">

    <RelativeLayout
        android:layout_width="fill_parent"
        android:layout_height="wrap_content"
        android:id="@+id/relativeVideoLayoutView"
        android:layout_centerInParent="true">
    <VideoView
        android:id="@+id/videoView"
        android:layout_centerInParent="true"
        android:layout_height="176px"
        android:layout_width="144px"/>
    </RelativeLayout>

    <LinearLayout
        android:layout_width="wrap_content"
        android:layout_height="wrap_content"
        android:orientation="horizontal"
```

```xml
        android:layout_centerHorizontal="true"
        android:layout_below="@+id/relativeVideoLayoutView">
    <ImageButton
        android:id="@+id/playRecordingBtn"
        android:layout_width="wrap_content"
        android:layout_height="wrap_content"
        android:background="@drawable/play" />
    <ImageButton
        android:id="@+id/bgnBtn"
        android:layout_width="wrap_content"
        android:layout_height="wrap_content"
        android:background="@drawable/record"
        android:enabled="false" />
</LinearLayout>
</RelativeLayout>
```

录制视频与录制音频的步骤相似，在录制音频的基础上加上与视频相关的特殊步骤即可。

步骤 03 创建MediaRecorder对象。

MediaRecorder video_recorder = new MediaRecorder();

步骤 04 设置音频和视频源。

创建MediaRecorder对象后，需要设置音频和视频源。使用setAudioSource方法设置音频源，传入一个想要使用的音频源常量。（设置方法已在8.1.4节录制音频中介绍过，此处不再赘述）。设置视频源可以使用setVideoSource方法。可能的视频源的值是在MediaRecorder.VideoSource类中定义的，其中只包含CAMERA和DEFAULT两个常量。其实这两个常量表示的含义一样，都是指设备上的主摄像头。关键代码如下。

video_recorder.setVideoSource(MediaRecorder.VideoSource.DEFAULT);

步骤 05 设置输出格式。

设置音频和视频源之后，可以使用MediaRecorder的setOutputFormat方法设置输出格式，传入要使用的格式。例如：

video_recorder.setOutputFormat(MediaRecorder.OutputFormat.DEFAULT);

可能的格式由MediaRecorder.OutputFormat中的常量来定义。
- **DEFAULT**：默认的输出格式。默认的输出格式根据设备的不同而不同。
- **MPEG_4**：指定音频和视频被录制在一个MPEG_4格式的文件中，扩展名是".mp4"。MPEG_4文件通常包含H.264、H.263或MPEG-4 Part 2编码的视频，以及AAC或MP3编码的音频。MPEG_4文件广泛应用于许多其他在线视频技术或消费电子设备上。
- **THREE_GPP**：指定的音频和视频将被录制到一个3GP格式的文件中，扩展名是".3gp"。3GPP文件通常包含使用H.264、H.263或MPEG-4 Part 2编码的视频和使用AMR或AAC编码的音频。

步骤06 设置音频和视频编解码器。

设置输出格式之后，需要指定想要使用的音频和视频编解码器，可以使用MediaRecorder的setVideoEncoder方法设置视频编解码器。

```
video.setVideoEncoder(MediaRecorder.VideoEncoder.DEFAULT);
```

可以使用的编解码器由MediaRecorder.VideoEncoder中的常量定义，具体常量如下。
- **DEFAULT**：默认的视频编解码器。多数情况下是H.263的编解码器，是Android设备上唯一支持的编解码器。
- **H263**：指定H.263为视频编解码器。H.263是1995年发布的编解码器，专门为低比特率视频传输而开发。它是许多早期Internet视频技术的基础，如Flash和RealPlayer早期都是使用该技术。
- **H264**：指定H.264为视频编解码器。H.264是当前先进的编解码器，被广泛应用于各种技术，如BlueRay、Flash。
- **MPEG_4_SP**：指定视频编解码器为MPEG-4 SP。MPEG-4 SP是MPEG-4 Part 2 Simple Profile，于1999年发布，为需要低比特率视频且不需要大处理器能力的技术而开发。

步骤07 设置音频和视频比特率。

使用MediaRecorder的setVideoEncodingBitRate方法设置视频编码比特率。视频的低比特率设置在256 000 bit/s（256 kbit/s）范围之内，而高比特率在3 000 000 bit/s（3 Mbit/s）范围之内。例如：

```
video_recorder.setVideoEncodingBitRate(150000);
```

使用MediaRecorder的setAudioEncodingBitRate方法设置音频编码比特率。8 000 bit/s是一个非常低的比特率，是适合在慢速网络上实时传输的音频。例如：

```
video_recorder.setAudioEncodingBitRate (8000);
```

步骤08 设置音频采样率。

音频采样率和编码比特率一样，它对于音频的质量也非常重要。可以使用MediaRecorder的setAudioSamplingRate方法设置音频采样率。采样率以Hz为单位，表示每秒采样的数量。采样率越高，在录制音频文件时可以表示的音频频率的范围就越大。8 000 Hz为低端的采样率，适合录制低质量的音频，48 000 Hz为高端的采样率，可用于DVD和其他高质量的音频格式。

video_recorder. setAudioSamplingRate (8000);

步骤 09 设置音频通道。

可以使用setAudioChannels方法指定将要录制的音频通道的数量。目前，大多数Android设备上，音频大都限制为单一通道麦克风，因此使用一个以上的通道不一定有好处。对于通道数量，一般是单声道为一个通道，而立体声为两个通道。例如：

video_recorder. setAudioChannels (1);

步骤 10 设置视频帧速率。

可以使用setVideoFrameRate方法控制每秒录制的视频帧数，通常每秒12~15帧的值足以表示运动。使用的实际帧速率最终取决于设备的能力。例如：

video_recorder. setVideoFrameRate(15);

步骤 11 设置视频大小。

可以通过setVideoSize方法设置宽度和高度值，以控制所录制视频的宽度和高度。视频的标准大小范围是176×144~640×480，许多设备甚至能支持更高的分辨率。例如：

video_recorder. setVideoSize (640, 480);

步骤 12 设置最大文件大小。

使用setMaxFileSize方法设置最大文件大小，单位为字节。例如：

video_recorder. setMaxFileSize (10000000); //约10 MB

为了确定是否已达到最大文件大小，需要在Activity中实现MediaRecorder.OnInfoListener，同时在MediaRecorder中注册它，然后系统会调用onInfo方法，检查其参数是否等于最大文件大小。

MediaRecorder.MEDIA_RECORDER_INFO_FILESIZE_REACHED

步骤 13 设置持续时间。

使用setMaxDuration方法设置最长持续时间，单位为毫秒。例如：

video_recorder. setMaxDuration(10000); //10 s

为了确定是否已达到最长持续时间，需要在Activity中实现MediaRecorder.OnInfoListener，同时在MediaRecorder中注册它。当达到最长持续时间时就会触发onInfo方法，检查其参数是否等于最长持续时间。

MediaRecorder.MEDIA_RECORDER_INFO_MAX_DURATION_REACHED

步骤 14 概要。

MediaRecorder有一个setProfile方法，接收CamcorderProfile实例作为参数。使用该方法允许根据预设值设置整个配置变量集。预设值中，CamcorderProfile.QUALITY_LOW指低质量视频捕获设置，CamcorderProfile.QUALITY_HIGH指高质量视频捕获设置。

步骤 15 输出文件。

使用setOutputFile方法设置输出文件的位置。例如：

video_recorder. setOutputFile（"/sdcard/video_recorded.mp4"）;

步骤 16 预览表面。

录制视频的过程中需要看到画面，因此需要为MediaRecorder指定一个取景器以预览要绘制的图像。另外，还需要使用SurfaceView和SurfaceHolder.Callback配合。

步骤 17 准备录制视频。

设置好MediaRecorder实例后，就可以使用prepare方法准备MediaRecorder了，代码如下。

video_recorder.prepare();

步骤 18 开始录制视频。

MediaRecorder实例准备好后，就可以开始录制视频了，代码如下。

video_recorder.start();

步骤 19 停止录制视频。

录制视频的过程中，可以通过stop方法停止录制，代码如下。

video_recorder.stop();

步骤 20 释放资源。

最后，一定不要忘记释放占用的资源，代码如下。

video_recorder.release();

创建一个名为"VideoCam"的Activity类，实现录制视频的功能。代码如下。

```
public class VideoCam extends Activity implements SurfaceHolder.Callback {
    private MediaRecorder recorder = null;
    private static final String OUTPUT_FILE = "/sdcard/uatestvideo.mp4";
    private static final String TAG = "RecordVideo";
    private VideoView videoView = null;
    private ImageButton startBtn = null;
```

```java
private ImageButton playRecordingBtn = null;
private Boolean playing = false;
private Boolean recording = false;

public void onCreate(Bundle savedInstanceState) {
    super.onCreate(savedInstanceState);
    setContentView(R.layout.main);
    startBtn = (ImageButton) findViewById(R.id.bgnBtn);
    playRecordingBtn = (ImageButton) findViewById(R.id.playRecordingBtn);
    videoView = (VideoView)this.findViewById(R.id.videoView);
    final SurfaceHolder holder = videoView.getHolder();
    holder.addCallback(this);
    holder.setType(SurfaceHolder.SURFACE_TYPE_PUSH_BUFFERS);
    startBtn.setOnClickListener(new OnClickListener() {
      public void onClick(View view) {
        if(!VideoCam.this.recording & !VideoCam.this.playing)
        {
          try
          {
            beginRecording(holder);
            playing=false;
            recording=true;
            startBtn.setBackgroundResource(R.drawable.stop);
          } catch (Exception e) {
            Log.e(TAG, e.toString());
            e.printStackTrace();
          }
        }
        else if(VideoCam.this.recording)
        {
          try
          {
            stopRecording();
            playing = false;
            recording= false;
            startBtn.setBackgroundResource(R.drawable.play);
          }catch (Exception e) {
```

```
            Log.e(TAG, e.toString());
            e.printStackTrace();
        }
      }
    }
});

playRecordingBtn.setOnClickListener(new OnClickListener() {
    public void onClick(View view)
    {
        if(!VideoCam.this.playing & !VideoCam.this.recording)
        {
            try
            {
                playRecording();
                VideoCam.this.playing=true;
                VideoCam.this.recording=false;
                playRecordingBtn.setBackgroundResource (R.drawable.stop);
            } catch (Exception e) {
                Log.e(TAG, e.toString());
                e.printStackTrace();
            }
        }
        else if(VideoCam.this.playing)
        {
            try
            {
                stopPlayingRecording();
                VideoCam.this.playing = false;
                VideoCam.this.recording= false;
                playRecordingBtn.setBackgroundResource (R.drawable.play);
            }catch (Exception e) {
                Log.e(TAG, e.toString());
                e.printStackTrace();
            }
        }
    }
```

```java
    });
}

public void surfaceCreated(SurfaceHolder holder) {
    startBtn.setEnabled(true);
}

public void surfaceDestroyed(SurfaceHolder holder) {
}

public void surfaceChanged(SurfaceHolder holder, int format, int width, int height) {
    Log.v(TAG, "Width x Height = " + width + "x" + height);
}

private void playRecording() {
    MediaController mc = new MediaController(this);
    videoView.setMediaController(mc);
    videoView.setVideoPath(OUTPUT_FILE);
    videoView.start();
}

private void stopPlayingRecording() {
    videoView.stopPlayback();
}

private void stopRecording() throws Exception {
    if (recorder != null) {
        recorder.stop();
    }
}

protected void onDestroy() {
    super.onDestroy();
    if (recorder != null) {
        recorder.release();
    }
}
```

```java
private void beginRecording(SurfaceHolder holder) throws Exception
{
    if(recorder!=null)
    {
        recorder.stop();
        recorder.release();
    }
    File outFile = new File(OUTPUT_FILE);
    if(outFile.exists())
    {
        outFile.delete();
    }
    try {
        recorder = new MediaRecorder();
        recorder.setVideoSource(MediaRecorder.VideoSource.CAMERA);
        recorder.setAudioSource(MediaRecorder.AudioSource.MIC);
        recorder.setOutputFormat(MediaRecorder.OutputFormat.MPEG_4);
        recorder.setVideoSize(320, 240);
        recorder.setVideoFrameRate(15);
        recorder.setVideoEncoder(MediaRecorder.VideoEncoder.MPEG_4_SP);
        recorder.setAudioEncoder(MediaRecorder.AudioEncoder.AMR_NB);
        recorder.setMaxDuration(20000);
        recorder.setPreviewDisplay(holder.getSurface());
        recorder.setOutputFile(OUTPUT_FILE);
        recorder.prepare();
        recorder.start();
    }
    catch(Exception e) {
        Log.e(TAG, e.toString());
        e.printStackTrace();
    }
}
```

除了设置成员之外，首先要做的就是创建一个显示Surface，用于支持摄像头预览，然后创建重要的beginRecording方法。该方法用于确保所有准备工作都已就绪，可以开始录制视频。

录制视频前先确保摄像头可用，然后检测输出文件是否存在，若存在则删除，接着遵循录制视频的一般流程建立并设置MediaRecorder。

8.2 使用电话API

基于Android平台的智能设备，特别是智能手机，都支持在程序中使用电话的相关服务。本节将介绍如何使用电话API编写代码实现拨电话、发送SMS和接受SMS的功能。

8.2.1 拨打电话

拨打电话最简单的方法是使用Intent.ACTION_CALL调用Android内部的拨号应用。该方法是先调用拨号应用，然后把电话号码传给拨号应用，接着显示在拨号盘上，开始拨打。

另外，还可以使用Intent.ACTION_DIAL调用Android内部的拨号应用。该方式也是首先调用拨号应用，然后填充电话号码，接着显示在拨号盘上，但是不发起拨打电话的动作，而是等待用户点击拨号盘上的拨打按钮。下面的代码演示了如何使用Intent.ACTION_DIAL和Intent.ACTION_CALL实现拨打电话的功能。

```java
dialintent = (Button) findViewById(R.id.dialintent_button);
dialintent.setOnClickListener(new OnClickListener() {
    public void onClick(View v) {
        Intent intent = new Intent(Intent.ACTION_DIAL, Uri.parse("tel:" + NUMBER));
        startActivity(intent);
    }
});

callintent = (Button) findViewById(R.id.callintent_button);
callintent.setOnClickListener(new OnClickListener() {
    public void onClick(View v) {
        Intent intent = new Intent(Intent.ACTION_CALL, Uri.parse("tel:" + NUMBER));
        startActivity(intent);
    }
});
```

需要注意，使用Intent.ACTION_DIAL实现拨打电话不需要特殊的权限，但是使用Intent.ACTION_CALL实现拨打电话需要给应用程序添加CALL_PHONE的权限。

8.2.2 发送SMS

Android SDK给开发人员提供了发送和接收SMS的一系列API。

【案例8-5】：编写程序实现发送SMS。为了发送SMS，首先需要在应用程序的"manifest.xml"文件中添加android.permission.SEND_SMS权限，然后再使用android.telephony.SmsManager类。

步骤 01 创建一个Android项目，名为"SMSExample"，然后编辑其布局XML文件，在布局文件中添加如下内容。

```xml
<?xml version="1.0" encoding="utf-8"?>
<!-- This file is /res/layout/main.xml -->
<LinearLayout xmlns:android="http://schemas.android.com/apk/res/android"
    android:layout_width="fill_parent"
    android:layout_height="fill_parent"
    android:orientation="vertical" >

    <LinearLayout xmlns:android="http://schemas.android.com/apk/res/android"
        android:layout_width="fill_parent"
        android:layout_height="wrap_content"
        android:orientation="horizontal" >
        <TextView
            android:layout_width="wrap_content"
            android:layout_height="wrap_content"
            android:text="Destination Address:" />
        <EditText
            android:id="@+id/addrEditText"
            android:layout_width="fill_parent"
            android:layout_height="wrap_content"
            android:phoneNumber="true"
            android:text="9045551212" />
    </LinearLayout>

    <LinearLayout xmlns:android="http://schemas.android.com/apk/res/android"
        android:layout_width="fill_parent"
        android:layout_height="wrap_content"
        android:orientation="vertical" >
        <TextView
            android:layout_width="wrap_content"
            android:layout_height="wrap_content"
            android:text="Text Message:" />
        <EditText
            android:id="@+id/msgEditText"
            android:layout_width="fill_parent"
            android:layout_height="wrap_content"
            android:text="hello sms" />
```

```xml
</LinearLayout>

<Button
    android:id="@+id/sendSmsBtn"
    android:layout_width="wrap_content"
    android:layout_height="wrap_content"
    android:onClick="doSend"
    android:text="Send Text Message" />

</LinearLayout>
```

步骤 02 新建一个名为"SendSMSActivity.java"的类文件，代码如下。

```java
import android.app.Activity;
import android.os.Bundle;
import android.telephony.SmsManager;
import android.view.View;
import android.widget.EditText;
import android.widget.Toast;

public class SendSMSActivity extends Activity
{
    @Override
    protected void onCreate(Bundle savedInstanceState) {
        super.onCreate(savedInstanceState);
        setContentView(R.layout.main);
    }

    public void doSend(View view) {
        EditText addrTxt = (EditText) findViewById(R.id.addrEditText);
        EditText msgTxt = (EditText) findViewById(R.id.msgEditText);
        try {
            sendSmsMessage(
                addrTxt.getText().toString(),
                msgTxt.getText().toString());
            Toast.makeText(this, "SMS Sent",
                Toast.LENGTH_LONG).show();
        } catch (Exception e) {
            Toast.makeText(this, "Failed to send SMS", Toast.LENGTH_LONG).show();
```

```
            e.printStackTrace();
        }
    }

    @Override
    protected void onDestroy() {
        super.onDestroy();
    }

    private void sendSmsMessage(String address,String message)throws Exception
    {
        SmsManager smsMgr = SmsManager.getDefault();
        smsMgr.sendTextMessage(address, null, message, null, null);
    }
}
```

界面布局XML文件中添加了两个文本编辑框：一个用于获取SMS接收者的地址（电话号码），另一个用于获取要发送的短消息。Android SDK给开发人员提供了发送和接收SMS的一系列API。除此之外，用户界面上还有一个按钮，用于触发发送SMS的动作。

代码中最重要的一个方法是sendSmsMessage方法。该方法使用SmsManager类的sendTextMessage方法发送SMS短消息。

可以在模拟器中测试SMS发送。首先启动该程序的两个实例，使用其中一个模拟器的端口号作为目的地址。该端口号就是出现在模拟器窗口标题栏的号码，它通常是类似"5554"这样的号码。点击发送按钮后，会看见一个通知消息出现在另一个模拟器上，表明SMS已经发送成功。

8.2.3 接收SMS

发送SMS之后，如何编写程序接收SMS呢？首先通过添加一个新的BroadcastReceiver派生类来监听android.provider.Telephony.SMS_RECEIVED这个动作。这个动作是当一个SMS短消息被设备接收后由Android平台自动广播出去的。在该动作上注册了自己的receiver后，每当SMS短消息收到后，应用就会被通知。

操作步骤如下。

步骤 01 向应用程序的"AndroidManifest.xml"文件中添加android.permission.RECEIVE_SMS权限，代码如下。

```
<?xml version="1.0" encoding="utf-8"?>
<!-- This file is AndroidManifest.xml -->
```

```xml
<manifest xmlns: android="http://schemas.android.com/apk/res/android"
  package="com.selfteaching.smsexample"
  android:versionCode="1"
  android:versionName="1.0" >

  <application
    android:icon="@drawable/icon"
    android:label="@string/app_name" >

    <activity
      android:name=".SMSExample"
      android:label="@string/app_name" >
      <intent-filter>
        <action android:name="android.intent.action.MAIN" />
        <category android:name="android.intent.category.LAUNCHER" />
      </intent-filter>
    </activity>

    <receiver android:name="SMSMonitorEx" >
      <intent-filter>
        <action android:name="android.provider.Telephony.SMS_RECEIVED" />
      </intent-filter>
    </receiver>
  </application>

  <uses-sdk android:minSdkVersion="4" />
  <uses-permission android:name="android.permission.SEND_SMS" />
  <uses-permission android:name="android.permission.RECEIVE_SMS" />
</manifest>
```

步骤 02 新建一个Android项目，命名为"SMSExample"，代码如下。

```java
import android.content.BroadcastReceiver;
import android.content.Context;
import android.content.Intent;
import android.telephony.SmsMessage;
import android.util.Log;
public class SMSMonitorEx extends BroadcastReceiver
```

```
{
    private static final String ACTION = "android.provider.Telephony.SMS_RECEIVED";

    @Override
    public void onReceive(Context context, Intent intent)
    {
        if(intent!=null && intent.getAction()!=null &&
    ACTION.compareToIgnoreCase(intent.getAction())==0)
        {
            Object[] pduArray= (Object[]) intent.getExtras().get("pdus");
            SmsMessage[] messages = new  SmsMessage[pduArray.length];
            for (int i = 0; i<pduArray.length; i++) {
                messages[i] = SmsMessage.createFromPdu( (byte[])pduArray [i]);
                Log.d("SMSMonitorEx", "From: " + messages[i].getOriginatingAddress());
                Log.d("MySMSMonitor", "Msg: " + messages[i].getMessageBody());
            }
            Log.d("MySMSMonitor","SMS Message Received.");
        }
    }
}
```

课后作业

一、填空题

1. Android录制音频最简单的方法是_____。

2. Android中MediaRecorder类是_____提供的录制音频和视频的功能类，使用它可以建立功能更加完善的_____，如可以控制录制的时长等。

3. MediaPlayer播放音频的最简单方式是_____。音频文件放置在应用程序的原始资源中。

4. _____是一个带有视频播放功能的视图，可以直接在布局中使用，使用起来非常简单。

5. 多媒体主要包括_____和_____。

二、选择题

1. 音频录制的主要API，（　　）是开始录制前准备MediaRecorder对象。

   A. MediaRecorder.setAudioSource()

   B. MediaRecorder.setOutputFormat()

   C. MediaRecorder.prepare()

   D. MediaRecorder.release()

2. 录制音频的过程中，（　　）个步骤是通过调用MediaRecorder.setOutputFile()方法完成的。

   A. 设置音频录制时采用的音频源（audio source）

   B. 指定输出音频数据存放的文件

   C. 设置输出音频数据的存储文件格式

   D. 停止录制

3. （　　）音频格式是一种新的音频压缩格式，类似于MP3的音乐格式。

   A. MP3

   B. AMR

   C. AAC

   D. OGG

4. MediaRecorder.VideoEncoder中的常量（　　）是为需要低比特率视频且不需要大处理器能力的技术而开发。

   A. MPEG_4_SP

   B. H264

   C. H263

   D. DEFAULT

三、操作题

利用Android Studio开发环境创建一个新的Android应用程序，实现音频的播放。要求从MediaStore中获取音频信息，并使用ListView显示；在点击歌曲时，实现该音频的播放功能，要求实现音频的顺序播放、单曲循环播放等功能。（提示：参照书中8.1.2节内容进行练习。）

第 9 章
综合应用：
基于桌面组件的开发

内容概要

第一次启动Android模拟器时，可以看到在桌面上有很多图标，点击这些图标，系统就会执行相应的程序。这些图标与PC机桌面上的快捷方式很像，但它又不完全是快捷方式，它还包括实时文件夹（Live Folder）和桌面组件（Widget）等，桌面组件应用既美观又方便用户操作。本章主要介绍桌面组件的开发，并将应用程序能轻松地放置到桌面上。

数字资源

【本章案例文件】："案例文件\第9章"目录下

微信扫码
- 配套资源
- 入门精讲
- 项目实战
- 日志记录

9.1 桌面快捷方式介绍

桌面组件快捷方式和PC机上的快捷方式一样,用于启动某一应用程序的某个组件(如Activity、Service等)。要在桌面上添加一个快捷方式很简单,只需长按桌面或者点击"Menu"按键(屏幕内三键导航模式下),就可以弹出添加桌面组件的选项(其中,Shortcuts对应"快捷方式";Widgets对应"窗口小部件",即用Widget开发的桌面插件;Folders对应"文件夹"),如图9-1左图所示,进入相应的选项后即可添加相应的桌面组件。

也可以在应用程序中通过代码将一个应用程序添加到"快捷方式"列表中:在"Home"界面点击"Menu"→"添加"→"快捷方式",会弹出如图9-1右图所示的界面,在其中可选择要创建的应用程序的快捷方式。

图 9-1 添加快捷方式

然后,在<intent-filter>里增加对应的<Action>,代码如下。

```xml
<activity android:label="@string/app_name" android:name=".ShortcutsActivity" >
    <intent-filter >
        <action android:name="android.intent.action.MAIN" />
        <action android:name="android.intent.action.CREATE_SHORTCUT"/>
        <category android:name="android.intent.category.LAUNCHER" />
    </intent-filter>
</activity>
```

通过上述配置,只有在触发增加快捷方式的这个动作时,才能找到程序的Activity,而程序快捷方式的名字、图标和事件等都可以在Activity里进行设置。本实例的Activity的代码如下。

```java
public class ShortcutsActivity extends Activity {
    /** Called when the activity is first created. */
    @Override
    public void onCreate(Bundle savedInstanceState) {
        super.onCreate(savedInstanceState);
        //要添加快捷方式的Intent
```

```
        Intent addShortcut;
        //判断是否要添加快捷方式
        if(getIntent().getAction().equals(Intent.ACTION_CREATE_SHORTCUT)){
            addShortcut = new Intent();
            //设置快捷方式的名字
            addShortcut.putExtra(Intent.EXTRA_SHORTCUT_NAME, "发送邮件");
            //构建快捷方式中专门的图标
            Parcelable icon = Intent.ShortcutIconResource.fromContext(this, R.drawable.mail_edit);
            //添加快捷方式图标
            addShortcut.putExtra(Intent.EXTRA_SHORTCUT_ICON_RESOURCE, icon);
            //构建快捷方式执行的Intent
            Intent mailtoIntent = new Intent(Intent.ACTION_SENDTO,Uri.parse("mailto:jercy@163.com" ));
            //添加快捷方式Intent
            addShortcut.putExtra(Intent.EXTRA_SHORTCUT_INTENT, mailtoIntent);
            setResult(RESULT_OK, addShortcut);
        }else{
            //取消
            setResult(RESULT_CANCELED);
        }
        finish();
    }
}
```

9.2 桌面组件——Widget

Widget可以提供一个full-featured apps的预览，例如，可以显示即将到来的日历事件，或者显示一首后台播放的歌曲的详细信息。当Widget被拖到桌面上时，会指定一个保留的空间来显示应用提供的自定义内容。用户可以通过这个Widget和应用交互，如暂停或切换歌曲等。如果有一个后台服务，可以按照自己的任务安排更新Widget，或者使用AppWidget框架提供一个自动更新机制。

9.2.1 AppWidget框架类

AppWidget的框架类主要包括4种。

（1）AppWidgetProvider：继承自BroadcastRecevier类，在AppWidget应用update、enable、disable和delete时接收通知；onUpdate和onReceive是最常用的方法，它们负责接收更新通知。

（2）AppWidgetProviderInfo：描述AppWidget的大小、更新频率和初始界面等信息，以XML文件形式存在于应用的"res\xml"目录下。

（3）AppWidgetManager：负责管理AppWidget，可向AppWidgetProvider发送通知。

（4）RemoteViews：可以在其他应用进程中运行的类，可向AppWidgetProvider发送通知。

其中，AppWidgetProvider类的主要方法有以下几种。

- **onDeleted(Context context, int[] appWidgetIds)**：删除。
- **onDisabled(Context context)**：禁用。
- **onEnabled(Context context)**：启用。
- **onUpdate(Context context, AppWidgetManager appWidgetManager, int[] appWidgetIds)**：更新。
- **onReceive(Context context, Intent intent)**：接收。

> **知识点拨**
>
> 因为AppWidgetProvider是继承自BroadcastReceiver类，所以可以重载onReceive方法，但必须在后台注册Receiver。

AppWidgetManager类可实现的主要方法有以下几种。

- **bindAppWidgetId(int appWidgetId, ComponentName provider)**：通过给定的ComponentName 绑定appWidgetId。
- **getAppWidgetIds(ComponentName provider)**：通过给定的ComponentName获取appWidgetId。
- **getAppWidgetInfo(int appWidgetId)**：通过appWidgetId获取AppWidget信息。
- **getInstalledProviders()**：返回一个List<AppWidgetProviderInfo>的信息。
- **getInstance(Context context)**：获取AppWidgetManager实例使用的上下文对象。
- **updateAppWidget(ComponentName provider, RemoteViews views)**：通过ComponentName对传进来的RemoteView进行修改，并刷新AppWidget组件。
- **updateAppWidget(int appWidgetId, RemoteViews views)**：通过appWidgetId对传进来的RemoteView进行修改，并刷新AppWidget组件。

9.2.2 AppWidget的简单例子：Hello AppWidget

本节用一个简单的例子进行演示，然后再针对这个例子讲解AppWidget的技术实现。操作步骤如下。

步骤01 新建一个项目"HelloAppWidget"，注意创建时可以不选Create Activity，如图9-2所示。

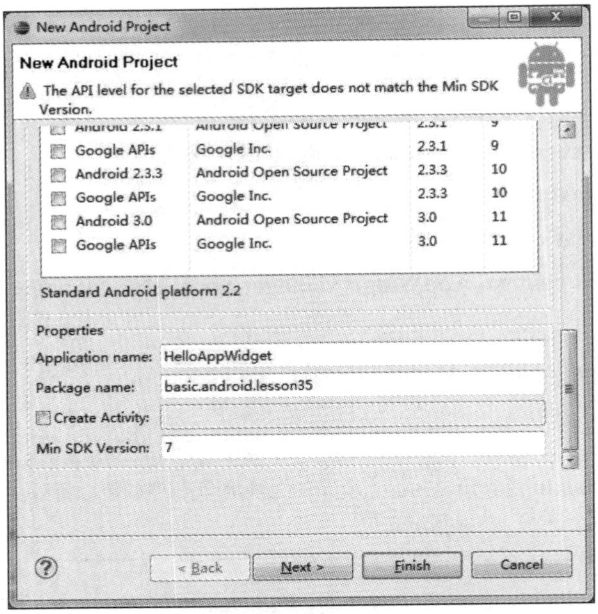

图 9-2 新建项目

步骤 02 新建一个Widget的显示布局文件"layout\widget.xml",代码如下。

```
<?xml version="1.0" encoding="utf-8"?>
<linearlayout xmlns:android="http://schemas.android.com/apk/res/android"
  android:layout_height="fill_parent"
  android:layout_width="fill_parent"
  android:orientation="vertical"
  android:gravity="center">

  <textview
    android:layout_height="wrap_content"
    android:layout_width="wrap_content"
    android:id="@+id/textView1"
    android:text="欢迎进入AppWidget的世界!"
    android:textcolor="#ff0000ff">
  </textview>

</linearlayout>
```

步骤 03 新建一个Widget的配置文件"xml\provider_info.xml",该文件用于提供Widget可以占用的屏幕长、宽、更新频率和所显示的布局文件(就是步骤02中创建的那个布局文件)等,代码如下。

```xml
<?xml version="1.0" encoding="utf-8"?>
<!-- appwidget-provider Widget的配置文件 -->
<!-- android:minWidth 最小宽度 -->
<!-- android:minHeight 最小高度 -->
<!-- android:updatePeriodMillis 组件更新频率（毫秒） -->
<!-- android:initialLayout 组件布局XML的位置 -->
<!-- android:configure Widget设置用Activity -->
<appwidget -provider="" xmlns:android="http://schemas.android.com/apk/res/android"
    android:initiallayout="@layout/widget"
    android:updateperiodmillis="86400000"
    android:minheight="72dp"
    android:minwidth="294dp">
</appwidget>
```

步骤 04 建立一个处理Widget请求的文件"HelloWidgetProvider.java"，它继承了AppWidgetProvider类，代码如下。

```java
import android.appwidget.AppWidgetManager;
import android.appwidget.AppWidgetProvider;
import android.content.Context;
import android.content.Intent;
import android.util.Log;

// AppWidgetProvider是BroadcastReceiver的子类，本质是个广播接收器，它专门用来接收来自
//Widget组件的各种请求（用Intent传递过来），所以如果让我给它起名的话，我会给它命名为
//"AppWidgetReceiver"，每一个Widget都要有一个AppWidgetProvider
public class HelloWidgetProvider extends AppWidgetProvider {
//每个请求都会传递给onReceive方法，该方法根据Intent参数中的action类型来决定自己处理还是分
//发给下面4个特殊的方法
    @Override
    public void onReceive(Context context, Intent intent) {
        Log.i("yao", "HelloWidgetProvider --> onReceive");
        super.onReceive(context, intent);
    }

//如果Widget自动更新的时间到了，或者其他会导致Widget发生变化的事件发生，或者Intent的
//值是android.appwidget.action.APPWIDGET_UPDATE，则会调用onUpdate方法。其他3个方法类似
    @Override
```

```java
    public void onUpdate(Context context, AppWidgetManager appWidgetManager, int[] appWidgetIds) {
//AppWidgetManager是AppWidget的管理器。appWidgetIds: 桌面上所有的Widget都会被分配一个
//唯一的ID标识，那么这个数组就是它们的列表
        Log.i("yao", "HelloWidgetProvider --> onUpdate");
        super.onUpdate(context, appWidgetManager, appWidgetIds);
    }


//当一个AppWidget从桌面上被删除时调用
    @Override
    public void onDeleted(Context context, int[] appWidgetIds) {
        Log.i("yao", "HelloWidgetProvider --> onDeleted");
        super.onDeleted(context, appWidgetIds);
    }


//当一个AppWidget第一次被放在桌面上时调用（同一个AppWidget可以放在桌面上多次，所以会//有这个说法）
    @Override
    public void onEnabled(Context context) {
        Log.i("yao", "HelloWidgetProvider --> onEnabled");
        super.onEnabled(context);
    }


//当一个AppWidget的最后一个实例从桌面上移除时会调用该方法
    @Override
    public void onDisabled(Context context) {
        Log.i("yao", "HelloWidgetProvider --> onDisabled");
        super.onDisabled(context);
    }

}
```

步骤 05 编辑"AndroidManifest.xml"文件，增加一个receiver标签，代码如下。

```xml
<?xml version="1.0" encoding="utf-8"?>
<manifest xmlns:android="http://schemas.android.com/apk/res/android" android:versionname="1.0"
android:versioncode="1" package="basic.android.lesson35">
  <uses-sdk "" android:minsdkversion="7">
  <application android:icon="@drawable/icon" android:label="@string/app_name">
```

```xml
<!-- receiver的android:name指向的是widget的请求处理器或者说请求接收者 -->
<receiver android:label="Hello,AppWidget" android:name=".HelloWidgetProvider">
    <intent -filter="">
        <!-- widget默认的事件action -->
        <action android:name="android.appwidget.action.APPWIDGET_UPDATE"></action>
    </intent>
    <!-- widget元数据，name是写死的，resource指的是widget的配置文件 -->
    <meta -data="" android:name="android.appwidget.provider" android:resource="@xml/provider_info">
</receiver>
</application>
</uses>
</manifest>
```

步骤 06 编译并运行程序。

Widget程序即使安装完毕也不会在程序列表中出现，因为它没有MainActivity。

下面介绍如何把Widget放到桌面上。操作步骤如下。

步骤 01 在模拟器的桌面上长按，弹出如图9-3所示的模拟器界面。

步骤 02 选择"窗口小部件"选项，如图9-4所示。

图9-3 模拟器界面

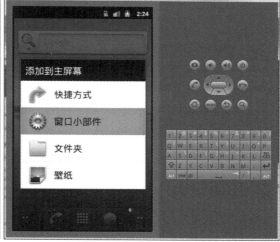

图9-4 选择"窗口小部件"选项

步骤 03 选择"Hello,AppWidget"选项，如图9-5所示。

步骤 04 当桌面上出现"欢迎进入AppWidget的世界！"的蓝色小字，则进入了AppWidget的世界，最终效果如图9-6所示。

图 9-5　程序出现在窗口小部件列表中　　　　图 9-6　最终效果

9.3　桌面天气预报程序设计

【案例】：开发一个可以获取天气预报信息的桌面小部件，并能实时异步更新界面。

操作步骤如下。

步骤 01 新建 "res\layout\weather_widget_layout.xml" 文件，用于描述部件的布局，代码如下。

```
<RelativeLayout xmlns:android="http://schemas.android.com/apk/res/android"
  android:layout_width="match_parent"
  android:layout_height="match_parent"
  android:background="@drawable/weather_widget_back">
  <TextView
    android:layout_width="wrap_content"
    android:layout_height="wrap_content"
    android:id="@+id/city_name"
    android:text="城市名称"
    android:textSize="20sp"
    android:layout_alignParentLeft="true"
    android:layout_marginTop="10dp"
    android:layout_marginLeft="15dp"/>
  <TextView
    android:layout_width="wrap_content"
    android:layout_height="wrap_content"
    android:id="@+id/cloud"
```

```xml
        android:text="风向"
        android:textSize="20sp"
        android:layout_marginTop="10dp"
        android:layout_marginRight="15dp"
        android:layout_alignParentRight="true"/>
    <LinearLayout
        android:layout_below="@id/city_name"
        android:layout_width="match_parent"
        android:layout_height="match_parent"
        android:orientation="horizontal"
        android:layout_marginBottom="10dp">
        <ImageView
            android:id="@+id/weather"
            android:layout_width="0dp"
            android:layout_height="match_parent"
            android:src="@drawable/little_rain"
            android:layout_weight="1"/>
        <TextView
            android:id="@+id/cur_temp"
            android:layout_width="0dp"
            android:layout_height="match_parent"
            android:layout_weight="1"
            android:gravity="center"
            android:text="20"
            android:textSize="50sp"/>
        <LinearLayout
            android:layout_height="match_parent"
            android:layout_width="0dp"
            android:layout_weight="1"
            android:gravity="center"
            android:orientation="vertical">
            <TextView
                android:id="@+id/low_temp"
                android:layout_width="match_parent"
                android:layout_height="wrap_content"
                android:text="最低 13"
                android:textSize="30sp"/>
            <TextView
```

```xml
            android:id="@+id/high_temp"
            android:layout_width="match_parent"
            android:layout_height="wrap_content"
            android:text="最高 20"
            android:textSize="30sp"/>
    </LinearLayout>
  </LinearLayout>
</RelativeLayout>
```

步骤 02 新建一个小部件的内容提供文件"res\xml\weather_info.xml",代码如下。

```xml
<appwidget-provider xmlns:android="http://schemas.android.com/apk/res/android"
    android:minWidth="214dp"
    android:minHeight="142dp"
    android:updatePeriodMillis="1000"
    android:initialLayout="@layout/weather_widget_layout" >
</appwidget-provider>
```

步骤 03 新建一个继承AppWidgetProvider类的子类WeatherWidget,代码如下。

```java
public class WeatherWidget extends AppWidgetProvider {
    public static WeatherFm wf = new WeatherFm();

    @Override
    public void onUpdate(Context context, AppWidgetManager appWidgetManager,
        int[] appWidgetIds) {
      super.onUpdate(context, appWidgetManager, appWidgetIds);
      Log.v("totoro","totoro1:" + appWidgetIds.length);
      LoadWeatherService.appWidgetIds = appWidgetIds;
      Log.v("totoro","totoro2:" + LoadWeatherService.appWidgetIds.length);
      context.startService(new Intent(context,LoadWeatherService.class));
    }

    public static RemoteViews updateRemoteViews(Context context) {
      RemoteViews view = new RemoteViews(context.getPackageName(),R.layout.weather_widget_layout);
      if (null == wf) {
        return null;
      } else {
        view.setTextViewText(R.id.cur_temp, wf.getCurTemperature());
```

```
            view.setTextViewText(R.id.low_temp, "低" + wf.getLowestTemperature());
            view.setTextViewText(R.id.high_temp, "高" + wf.getHighestTemperature());
            return view;
        }
    }
}
```

步骤 04 在"AndroidManifest.xml"文件中注册，代码如下。

```
<receiver
    android:name="com.monde.mondewidget.weatherwidget.WeatherWidget"
    android:icon="@drawable/ic_weather_widget">
    <intent-filter>
        <action android:name="android.appwidget.action.APPWIDGET_UPDATE"/>
    </intent-filter>
    <meta-data
        android:name="android.appwidget.provider"
        android:resource="@xml/weather_info"/>
</receiver>
```

以上是创建天气预报Widget的基本步骤，要实现天气预报内容的实时更新，还需要做更多的工作。图9-7中列出了这个桌面部件要用到的所有类。因为涉及到网络访问，所以在WeatherWidget类的onReceive方法中，无法直接对UI进行更新。

图 9-7　实例中所需的类

在步骤03中的context.startService(new Intent(context,LoadWeatherService.class))是代码的关键部分，用以实现启动网络访问、获取天气预报信息的功能。LoadWeatherService类的代码如下。

```
public class LoadWeatherService extends Service implements Runnable{
    private static Object isLock = new Object();
    private static boolean isThreadRun = false;
    public static int[] appWidgetIds;
```

```java
    @Override
    public int onStartCommand(Intent intent, int flags, int startId) {
        Log.v("onStartCommand","totoro:" + intent.toString());
        new Thread(this).start();
//      synchronized (isLock) {
//          if (!isThreadRun) {
//              isThreadRun = true;
//              new Thread(this).start();
//          }
//      }
        return super.onStartCommand(intent, flags, startId);
    }

    @Override
    public IBinder onBind(Intent intent) {
        // TODO Auto-generated method stub
        return null;
    }

    @Override
    public void run() {
        Looper.prepare();
        Log.v("onStartCommand","totoro");
        BDLocationUtils utils = new BDLocationUtils(this.getApplicationContext());
        utils.requestBDLocation();
        AppWidgetManager manager = AppWidgetManager.getInstance(this);
        WeatherQueryImpl impl = new WeatherQueryImpl(this);
        WeatherWidget.wf = impl.weatherQuery(utils.cityCode);
        RemoteViews view = WeatherWidget.updateRemoteViews(this);
        if (null != view) {
            manager.updateAppWidget(appWidgetIds, view);
        } else {
            Log.e("run", "更新失败");
        }
        stopSelf();
        Looper.loop();
    }
```

在这个服务中，启动了一个线程去处理网络访问，获取天气信息并使用返回的数据更新了UI，实际的天气预报获取操作均在WeatherQueryImpl类中实现，代码如下。

```java
public class WeatherQueryImpl implements WeatherQuery {
    private Context context;

    public WeatherQueryImpl(Context context) {
        this.context = context;
    }

    @Override
    public WeatherFm weatherQuery(String cityCode) {
        Log.v("weatherQuery","totoro:" + cityCode);
        String URL1 = "http://www.weather.com.cn/data/sk/" + cityCode + ".html";
        String URL2 = "http://www.weather.com.cn/data/cityinfo/" + cityCode + ".html";
        WeatherFm wf = new WeatherFm();
        wf.setCityCode(cityCode);
        String Weather_Result = "";
        HttpGet httpRequest = new HttpGet(URL1);
        // 获得当前温度
        try {
            HttpClient httpClient = new DefaultHttpClient();
            HttpResponse httpResponse = httpClient.execute(httpRequest);
            if (httpResponse.getStatusLine().getStatusCode() == HttpStatus.SC_OK) {
                // 取得返回的数据
                Weather_Result = EntityUtils.toString(httpResponse.getEntity());
                Log.v("totoro",Weather_Result);
            }
        } catch (Exception e) {
            e.printStackTrace();
            return null;
        }
        //以下是对返回JSON数据的解析
        if(null != Weather_Result&&!"".equals(Weather_Result)){
            try {
                JSONObject JO = new JSONObject(Weather_Result).getJSONObject("weatherinfo");
                wf.setCurTemperature(JO.getString("temp"));
                wf.setCityName(JO.getString("city"));
```

```java
        } catch (JSONException e) {
            e.printStackTrace();
            return null;
        }
    }

    Weather_Result = "";
    httpRequest = new HttpGet(URL2);
    // 获得HttpResponse对象
    try {
        HttpClient httpClient = new DefaultHttpClient();
        HttpResponse httpResponse = httpClient.execute(httpRequest);
        if (httpResponse.getStatusLine().getStatusCode() == HttpStatus.SC_OK) {
            // 取得返回的数据
            Weather_Result = EntityUtils.toString(httpResponse.getEntity());
            Log.v("totoro",Weather_Result);
        }
    } catch (Exception e) {
        e.printStackTrace();
        return null;
    }
    //以下是对返回JSON数据的解析
    if(null != Weather_Result&&!"".equals(Weather_Result)){
        try {
            JSONObject JO = new JSONObject(Weather_Result).getJSONObject("weatherinfo");
            wf.setWeatherCondition(JO.getString("weather"));
            wf.setLowestTemperature(JO.getString("temp2"));
            wf.setHighestTemperature(JO.getString("temp1"));
        } catch (JSONException e) {
            e.printStackTrace();
            return null;
        }
    }
    return wf;
}

@Override
```

```
    public String cityQuery() {
        try {
            return LocationUtils.getCNByGPSlocation(context);
            //return LocationUtils.getCNByWIFILocation(context);
        } catch (Exception e) {
            e.printStackTrace();
            return null;
        }
    }

    public WeatherFm getLocalWeather () {
        return weatherQuery(LocationCode.CHINESE_LOCAL_CODE.get(cityQuery()));
    }
```

代码中联网获取天气预报信息的网站是中国天气网（http://www.weather.com.cn/），它是比较权威的。例如，在此网站中选用如下接口：

http://www.weather.com.cn/data/sk/101280601.html

该接口返回的数据格式如下。

```
"weatherinfo": {
    "city": "深圳",
    "cityid": "101280601",
    "temp": "31",
    "WD": "东南风",
    "WS": "3级",
    "SD": "58%",
    "WSE": "3",
    "time": "17:10",
    "isRadar": "1",
    "Radar": "JC_RADAR_AZ9755_JB"
}
```

参考文献

[1] 马西卡诺. Android编程权威指南[M]. 4版. 王明发,译. 北京: 人民邮电出版社, 2021.

[2] 蒲晓妮,赵睿. Android应用程序开发基础[M]. 北京: 化学工业出版社, 2022.

[3] 启舰. Android自定义控件高级进阶与精彩实例[M]. 北京: 电子工业出版社, 2020.

[4] 明日学院. Android开发从入门到精通:项目案例版[M]. 北京: 中国水利水电出版社, 2017.

[5] 郭霖. 第一行代码Android[M]. 3版. 北京: 人民邮电出版社, 2020.

[6] 王向辉,张国印,沈洁. Android应用程序开发[M]. 3版. 北京: 清华大学出版社, 2016.